RADON

A Homeowner's Guide to Detection and Control

RADON

A Homeowner's Guide to Detection and Control

Bernard L. Cohen, D.Sc.
with Drew Nelson
and the Editors of Consumer Reports Books

Consumers Union
Mount Vernon, New York

Published by arrangement with Smith & Kraus, Inc.

Library of Congress Cataloging-in-Publication Data
Cohen, Bernard Leonard, 1924–
Radon : a homeowner's guide to detection and control.

Includes index.
1. Radon—Toxicology. 2. Radon. 3. Radiation dosimetry.
4. Air—Pollution, Indoor—Hygienic aspects.
5. Housing and health. 6. Consumer education. I. Consumer Reports Books. II. Title.
RA1247.R33C63 1987 613′.5 87-71010
ISBN 0-89043-227-9
ISBN 0-89043-080-2 (pbk.)

First printing, August 1987
Manufactured in the United States of America

Radon: A Homeowner's Guide to Detection and Control is a Consumer Reports Book published by Consumers Union, the nonprofit organization that publishes *Consumer Reports*, the monthly magazine of test reports, product Ratings, and buying guidance. Established in 1936, Consumers Union is chartered under the Not-For-Profit Corporation Law of the State of New York.

The purposes of Consumers Union, as stated in its charter, are to provide consumers with information and counsel on consumer goods and services, to give information on all matters relating to the expenditure of the family income, and to initiate and to cooperate with individual and group efforts seeking to create and maintain decent living standards.

Consumers Union derives its income solely from the sale of *Consumer Reports* and other publications. In addition, expenses of occasional public service efforts may be met, in part, by nonrestrictive, noncommercial contributions, grants, and fees. Consumers Union accepts no advertising or product samples and is not beholden in any way to any commercial interest. Its Ratings and reports are solely for the use of the readers of its publications. Neither the Ratings nor the reports nor any Consumers Union publications, including this book, may be used in advertising or for any commercial purpose. Consumers Union will take all steps open to it to prevent such uses of its materials, its name, or the name of *Consumer Reports*.

Contents

Preface

Some readers may be interested in knowing something about the author and his involvement in the radon problem. I am a 62-year-old professor of Physics and of Radiation Health at the University of Pittsburgh. I have been continuously involved in radiation research since 1947. After receiving my doctorate in 1950, I spent eight years at Oak Ridge National Laboratory in Tennessee, where I was a group leader in charge of cyclotron research. Since 1958 I have been continuously at the University of Pittsburgh except for serving as a visiting scientist for periods of one to nine months at a number of laboratories, including Argonne, Los Alamos, and Brookhaven National Laboratories, Stanford University, and the Institute for Energy Analysis. From 1965 to 1978 I served as Director of the University of Pittsburgh accelerator laboratory.

Since 1973, my principal research field has been the health effects of radiation. I first became interested in radon in 1975 in connection with its health impacts from uranium mining and milling. I began my first research project on radon in homes in 1977. In 1981, I set up a laboratory for developing improved techniques of measuring radon in homes, and my first pilot survey, covering 169 homes in the Pittsburgh area, commenced in 1982. By 1984, I had developed the measurement technique I was striving for to the point where I could use it in a pilot survey. After further technique development, I began a one-year national survey of a few thousand houses in late spring of 1985.

In June 1985, I was invited to speak to a group of citizens worried about radon in Bethlehem, Pennsylvania. The only commercially available measurement at that time cost $50, and the vendor was recommending three units for each house, which made it hardly affordable for most people. Feeling the pressure from the audience during the discussion period, on the spur of the moment I blurted out that I would provide a measurement to anyone who contributed $10 to my research program. There were reporters in the audience, and the word spread. By September, I realized that a more formal arrangement would be necessary. I also understood that I could derive a great deal of useful research information from these measurements. At about that time, government support for university research on radon in homes was being cut back, leaving me in need of research funding.

I therefore set up a regular radon measurement service under university auspices. We raised the price to $12, largely to provide support for the research program, but also to make it easy for private enterprise to compete. I have always considered it an important part of my mission to encourage private enterprise to do radon measurements in homes. I have made our techniques available to all, and have offered to serve as a free consultant. Well over a hundred prospective entrepreneurs have taken me up on this, and dozens of them have set up businesses specializing in radon measurement. I am convinced that private enterprise can provide our measurement service for well below our $12 price. Since we do no marketing, I expect them to "put us out of business."

In the meantime, we continue to make measurements and gather data. As of this writing, in January 1987, we have completed more than 52,000 measurements and written five research papers with the data collected.

Enough . . . let us begin.

Bernard L. Cohen

Introducing Radon

Uranium is a naturally occurring element found in all rock and soil, at least in trace amounts. Unlike most other elements, however, a uranium atom does not last "forever"; rather, it has a small probability of changing into an atom of another element, thorium. In this process an atomic "bullet," called an *alpha particle*, is shot out at a speed of about 10,000 miles per second, fast enough to go around the world in less than three seconds. High-speed atomic particles like this are called *radiation*.

The probability for an individual uranium atom to undergo this transformation is very small, only one chance in 7 billion per year. Since the Earth is only around 4 billion years old, about half of the uranium atoms originally on the planet have not yet experienced it; that is why they are still here. However, as small as the probability is for an individual atom to transform, the number of uranium atoms in a pound of ordinary rock or soil is so large that about a thousand of them transform every minute, shooting out radiation in the process. This material is therefore active in emitting radiation, and is consequently called *radioactive*.

Uranium is only one of several elements that are naturally radioactive. One well-known naturally radioactive element is potassium, which is necessary for life. Carbon, the key element in all molecules of living things, is also normally radioactive, though very weakly. Our bodies, as well as rock and soil, are therefore radioactive. Radioactivity is, and always has been, a part of nature.

When a uranium atom transforms into a thorium atom, the release of radiation means that the atom itself loses energy. It is therefore said to *decay*. Half of all uranium atoms in a given sample will decay in 4.5 billion years. Uranium is therefore said to decay with a *half-life* of 4.5 billion years.

The thorium atom into which the uranium atom decays does not survive for very long because it also undergoes radioactive decay, as does its decay product, as does the product of that decay, and so on—until after five successive decay processes extending over a few hundred thousand years, it becomes an atom of radium.

These radioactive decay processes do not move the atom physically. For the most part, radium atoms are at the same place as the original uranium atoms from which they are descended. But this situation changes when the radium atom decays.

Radium decays with a half-life of 1,600 years into another radioactive element called *radon*. Unlike all its predecessors, radon is a gas and is therefore free to move about. If it is formed in soil within a few feet of the surface, it has a good chance of percolating up out of the ground into the air. As an average over the world, about six atoms of radon emerge from every square inch of soil every second.

Once in the atmosphere, radon mixes with other components of air, and diffuses up to high altitudes, thus becoming very much diluted. Since radon is heavier than most other gases in the air, it doesn't rise quite as high; most air components rise easily to an altitude of about four miles, but radon atoms mostly stay below one mile in altitude. Near ground level, there is typically one radon atom for each *10 quintillion* atoms of air (1/10,000,000,000,000,000,000).

If the radon happens to emerge from the ground beneath a house and if there is an easy pathway for it to enter, the situation is somewhat different. Instead of becoming distributed up to a height of a mile, it is trapped within the house and can only diffuse up to the height of the roof, perhaps 30 feet. It is therefore much less diluted, making radon levels inside houses typically ten times higher than outdoors.

Radon has a half-life of 3.8 days. Its decay with emission of an alpha particle is followed within about an hour by a series of four further decays, two of them accompanied by emission of alpha particles and the other two accompanied by other types of radiation. The short-lived atoms into which a radon atom decays are actually isotopes of polonium, lead, and bismuth, but they are referred to collectively as *radon daughters*, or, by those sensitive to questions of gender, as *radon progeny*. The radon daughter atoms float around in the air during their few minutes of existence, often becoming attached to dust particles.

In summary, a radon atom in the air decays within a few days into its short-half-life radon daughters, which decay within about an hour; with these decays, three alpha particles are emitted, one by radon and two by its daughters.

Before proceeding, it is useful to compare alpha particles with other radiation. There are many types of radiation in nature, but they fall into two classes, *ionizing* and *non-ionizing*. Ionizing radiation is sufficiently energetic to knock electrons loose from molecules, which gives a new dimension to the damage they can do. Non-ionizing radiation, such as radio, TV, microwave, infrared, and visible light, generally does not have this ability. We are concerned here only with ionizing radiation. It occurs in many forms, including ultraviolet, X rays, neutrons, and protons, but the types normally emitted in radioactive decay are *alpha*, *beta*, and *gamma* rays. These three differ markedly in their ability to penetrate materials. Gamma rays, which are high-energy X rays, can penetrate to any part of the human body, while beta particles which are high-speed electrons, can penetrate only about ⅛-inch of solid material. Alpha particles are much heavier than betas and move much slower, even though they have typically ten times more energy. They can penetrate only through one sheet of paper.

One might think that the alpha particles shot out by radon and its daughters as they float around in the air give us an appreciable dose of radiation, but that is not the case. Gamma rays and beta particles, from cosmic rays coming in from outer space, from radioactive materials in the ground and in building materials, and from the radioactive potassium in our bodies, give us much more radiation exposure. In fact, about 15,000 atomic bullets from these sources strike our bodies every second. They are also very penetrating, easily passing through the air and getting deep into our bodies. The alpha particles from radon, on the other hand, are much fewer in number and they are much less penetrating, being stopped by about one inch of air. They cannot get through our clothing, and if they strike unclothed parts of our bodies they can seldom get through the outer dead layers of skin. One would think that radon was the least of our radiation problems.

That would be so except that when we breathe, we are constantly passing air into our lungs and out of them. In this process, the radon gas simply goes in and out, doing little damage, but the radon daughters, being basically solid materials, and sometimes being electrically charged, can stick to the surfaces of our bronchial tubes. This puts them right where they can do the most harm, for the cells lining our bronchial tubes are among the cells of our body most sensitive to radiation-induced cancer. The alpha particles emitted in the decay of radon daughters, in spite of their poor penetrating power, can reach these very sensitive cells because they are deposited so close to them. To make matters very much worse, alpha particles are much more efficient than other types of radiation for inducing cancer. The very fact that they are not penetrating means that they dump a lot of their energy into each of the biological cells they pass through, and this large release of energy into a single cell is just what is needed to initiate a cancer. As a result, an alpha particle is a hundred times more likely to cause cancer than other types of radiation, *if* it can reach the target cells. Our breathing processes allow the alpha particles from radon daughters to reach these cells.

As a result, radon is believed to be an important cause of lung cancer, killing about 10,000 Americans each year. Only cigarette smoking causes more lung cancer deaths per year. And in perhaps one out of a thousand American homes, radon levels are so high they pose a greater lung cancer risk than smoking a pack of cigarettes per day.

Radon in our homes gives the average American more radiation exposure than all other sources of radiation combined. People who live near the Three Mile Island nuclear power plant, for example, get more radiation exposure from radon in their homes *every day* than they received *in total* from the highly publicized accident there in 1979. People who lived very close to the Chernobyl nuclear power plant in the Soviet Union, and were not evacuated until 36 hours after the accident, still received less exposure from that accident than they can expect to receive over their lifetimes from radon in their homes. Some people worry about radiation exposure from diagnostic X rays, but Americans on the average get ten times as much radiation from radon as from medical and dental X rays, and 100 times as much from radon as from nuclear bomb test fallout. Compared to any other source of radiation, or even to all other sources combined, radon is the undisputed leader.

Life in the twentieth century exposes us to a multitude of risks—or at least we hear more about these risks than ever before. We can't worry about all of them. Should we worry about radon? There are two other questions we must ask to arrive at a rational answer: First, how big is the risk? Second, how difficult is it to reduce this risk? On the first point, one answer at least is clear. Since radon is by far our most important source of radiation exposure, anyone who worries about radiation in any form should be worried about radon in the home.

On the second point, at least the first step is easy—to obtain a measurement of the radon level in your home. In most cases, that will end your worry. If it doesn't, the rest of this book is for you. It reviews the present state of our knowledge on how to handle the radon problem.

2

The Lung Cancer Connection

Radiation Pioneers Discover a Miracle Technology

In November 1895, Professor Wilhelm Konrad Roentgen was experimenting with a gas discharge tube in his laboratory in the town of Würzburg, Bavaria, when he made a discovery that would forever change the face of science.

Roentgen was studying the luminous streams that appeared in a gas discharge tube when high-voltage electricity was applied to it. As part of this particular experiment, Roentgen wrapped heavy black paper around the device as a light barrier before electrifying it. As he switched on the power, Roentgen noticed a totally unexpected phenomenon. A crystal plate on a nearby workbench began to emit a curious glow. Distracted by this light, Roentgen switched off his gas tube to investigate. The glow ceased.

Repeating his actions, Roentgen observed that when the power to the tube was on, the plate glowed. When the power was off, the glow disappeared. He puzzled over the event and came to the conclusion that "rays" of some sort were crossing the open space between his gas tube and the crystals. Not knowing what he was dealing with, Roentgen applied the universal scientific symbol for the unknown quantity to his discovery and called them *X rays*.

In a series of follow-up experiments, Roentgen discovered one of the most useful properties of X rays—their ability to darken photographic plates. He also discovered that his X rays could penetrate most common substances to some degree. He illustrated this property by having his wife place her hand over a lightproofed photographic plate in the beam. The X rays left a perfect representation of her hand on the developed plate, including shadows of the soft tissues and excellent definition of the bones within, with the ring on her finger clearly visible.

The X rays had easily penetrated the soft tissues of her hand and, to lesser degrees, the bones and the metal of the ring. Roentgenography, or X-ray diagnosis, had thus been born. Roentgen's work amazed a world of physicians and scientists, who now had a means of seeing inside the human body without opening it surgically.

In a short time, however, the darker side of this miracle technology would reveal itself.

Uncontrolled Experiments Reveal the Dangers of Radiation

In Chicago, Emil H. Grubbe, following Roentgen's technique, began generating X rays using his own gas discharge tubes. During the course of his experiments, Grubbe overexposed and burned the fingers of one of his hands. The burns turned into dermatitis and led to the amputation of his fingers and, eventually, the entire hand.

Grubbe's misfortune with X rays opened a door for physicians who had been searching without luck for a cancer treatment. They reasoned that X rays, which could penetrate the body and destroy tissue, could be targeted against the cells of a malignancy. But they soon learned, as had Grubbe, that an ally as powerful as X rays could exact a toll if used without discretion.

In the early twentieth century, physicians discovered that cancers appeared with unusual frequency in patients given large doses of X rays. In fact, some of the physicians involved in such treatment of patients fell victim themselves to the cumulative effects of indirect X-ray exposure. They now began to suspect that X-ray radiation could cause cancer.

The Discovery of Radioactivity

Shortly after Roentgen's discovery of X rays, a French physicist, Henri Becquerel, investigated the possibility that phosphorescent materials might emit X rays as well as light after exposure to light. He placed various phosphorescent materials near photographic plates that were carefully wrapped in black paper, and looked for darkening of the plates from the X rays that might be emitted. He tried many phosphorescent materials, but the only one that darkened the photographic plate was a compound of uranium. He soon found that anything containing uranium, whether or not it was phosphorescent, or whether or not it had been exposed to light, would darken photographic plates. After further experimentation, he concluded that this was because uranium was emitting a radiation different from X rays. Becquerel had discovered radioactivity.

This discovery, in 1896, was quickly followed up by many other scientists, including, most notably, Pierre and Marie Curie, a married couple then living in France.

The Discovery of Radioactive Decay Products

Pierre, then a professor of physics at Municipal College in Paris, and Marie, a student at the Sorbonne, were together engaged in experiments measuring the electrical conductivity of air. In the course of this work, the Curies discovered that pitchblende, the mineral in which uranium is most commonly found, was four times more naturally radioactive than uranium itself. They reasoned that pitchblende must contain another radioactive material in addition to uranium.

Madame Curie devoted herself to the chemical separation of the substance that imparted this remarkable radioactivity to pitchblende. After endless tedious hours she was able to refine a substance that was many thousands of times more radioactive than pitchblende. She named her July 1898 discovery *polonium*, in honor of Poland, the country of her birth.

Fueled by this find, Madame Curie worked tirelessly with her husband and another collaborator toward a second stunning discovery in that same year—radium.

Becquerel accidentally learned of radium's danger as a result of carrying a small vial of the substance in his pocket. He suffered skin burns from the element's radioactive emanations. Madame Curie herself sustained many burns on the skin of her fingers and hands during her long hours of experimentation with radium. However, although she exposed herself in many ways—including the inhalation of radon gas—to what must have been enormous amounts of radiation, she lived well beyond the average life expectancy of her time, dying in 1934 of pernicious anemia (not related to radiation) at the age of 66. Those exposed to high levels of radiation do not necessarily suffer from the effects; they only have a higher *probability* of getting cancer than if they had not been exposed.

In the flurry of research following from the work of Becquerel and the Curies, several new radioactive elements were discovered. These occur in such small quantities that they could never be detected by normal chemical procedures, but the radiations they emitted made their presence obvious. One of these newly discovered elements was radon.

Radon and Lung Cancer

Emil Grubbe, Henri Becquerel, and Marie Curie were not the first persons to suffer the negative effects of working in radiation environments. At least as long ago as the sixteenth century, miners in central Europe were suffering from what we now know to be radon-induced lung cancers.

The earliest existing record of this health problem is in a sixteenth-century book on mining. It seems that many of the men who worked in underground silver mines in the Erz Mountains, which separated what is now East Germany from Bohemia (now Czechoslovakia), were dying from what was called at the time ***Bergkrankheit*** or "mountain sickness."

The mines continued to operate, and the miners continued to die from "mountain sickness" through the centuries. In the nineteenth century the mine hired physicians, and in 1879 two of them recognized that the disease was actually a cancer of the lung. Although one or two cases had been reported earlier in the medical literature, this was the first real recognition of lung cancer as an important disease. It is now killing 140,000 Americans each year.

Early in the twentieth century it was found that the Erz Mountain mines contained high levels of radon. This led to the hypothesis that it was the radon in the mines that was causing lung cancer in the miners, and by the 1940s this idea gained a great deal of acceptance.

Shortly thereafter, with the beginning of the uranium mining industry in the United States, regulations on exposure to radon were formulated. The maximum allowable average exposure was called one *working-level*, a term still used. But enforcement of the regulations was left to state mine inspectors, who knew all about coal-mine hazards but nothing about radon. As a result, little attention was paid to radon, and average exposures of 5, 10, or more working levels were common.

In the early 1950s, the U.S. Public Health Service started a study of these miners, in part to throw further light on the radon problem. The Erz Mountain mines were notorious for a wide variety of dangerous chemicals and cold, wet working conditions. Could these, rather than radon, be the cause of the lung cancer? In contrast, the U.S. mines were dry and generally free of dangerous chemicals. The one thing they had in common was a high level of radon.

Excess Lung Cancers Relate Directly to Radon Levels

By the early 1960s, the Public Health Service study began finding an excess of lung cancer among the miners. Within a few years, the data became unmistakable. The American uranium miners with radon exposure comparable to the Erz Mountain miners were found to suffer comparable rates of excess lung cancers. Those who experienced lower radon exposure registered proportionately lower excess lung cancer rates.

Because of these findings, strong action was taken to improve mine ventilation, and by 1970 the radon levels in U.S. uranium mines had been reduced dramatically. It is still too early to evaluate the results of improved ventilation in these mines because radiation-induced cancer has a ten-year latency period—an interval of ten or more years before symptoms appear. Miners continue to die of excess lung cancers from their radon exposure prior to 1970.

The principal data on the relationship between radon and lung cancer are shown in Fig. 2–1, which is a plot of risk per unit of exposure vs. exposure. In such a plot, if the risk is proportional to the exposure, the data will lie on a horizontal line. We see that the data on U.S. uranium miners deviate from a horizontal line only by a factor of 2, while the exposure varies by a factor of 100.

Much more data became available—from fluorspar mines in Newfoundland to zinc, lead, and iron mines in Sweden—and are shown in Fig. 2–1. Where radon levels were high, the miners working there were found to have excess lung cancers roughly in proportion to their radon exposures.

In 1976, a study of Czechoslovakian miners who began work

about 1950 was released. Radon levels were much lower than in earlier years, but again there was an excess of lung cancer. This excess was less than that experienced by the earlier miners, and the decrease roughly paralleled the decrease in radon levels. More recently, a study of Canadian miners has become available. As is evident in Fig. 2–1 their excess lung cancer rates were about the same as those for the U.S. and Czechoslovakian miners with the same radon exposure.

The lowest exposure in Fig. 2–1 is 80 WLM (a WLM is a working level month, the equivalent of one working level of exposure for one miner's working month, or 170 hours), which corresponds roughly to 6 pCi/l in one's home throughout life. But for the miners, the exposures were at a much higher rate—the 80 WLM was typically received in a five-year period, so exposure *rates* were fifteen times higher than for a person with 6 pCi/l in his home. It is not known whether total exposure is the only thing that counts, or whether exposure rate is also important.

There is abundant evidence, notably from studies of Japanese atomic bomb survivors and of a group of British patients treated with large doses of X rays for arthritis of the spine, that other types of radiation produce lung cancer. It was not a simple matter to relate these exposures to those from radon, but this has now been done. It turns out, as seen in Fig. 2–1, that the lung cancer risk per unit dose of radiation is the same.

For good geochemical reasons, uranium and coal do not occur together, and as a result, coal mines have very little radon. But coal mines do have lots of diesel fumes, dust, and other agents that might cause lung cancer. This provides another test of whether radon is the cause of lung cancer in uranium miners. Coal miners have high rates of nearly every respiratory disease, including bronchitis, pneumoconiosis, and pneumonia. At one time it was thought that they also suffered a high rate of lung cancer, but more careful studies have shown conclusively that lung cancer incidence among coal miners is very close to the

FIG. 2–1. LUNG CANCER IN MINERS

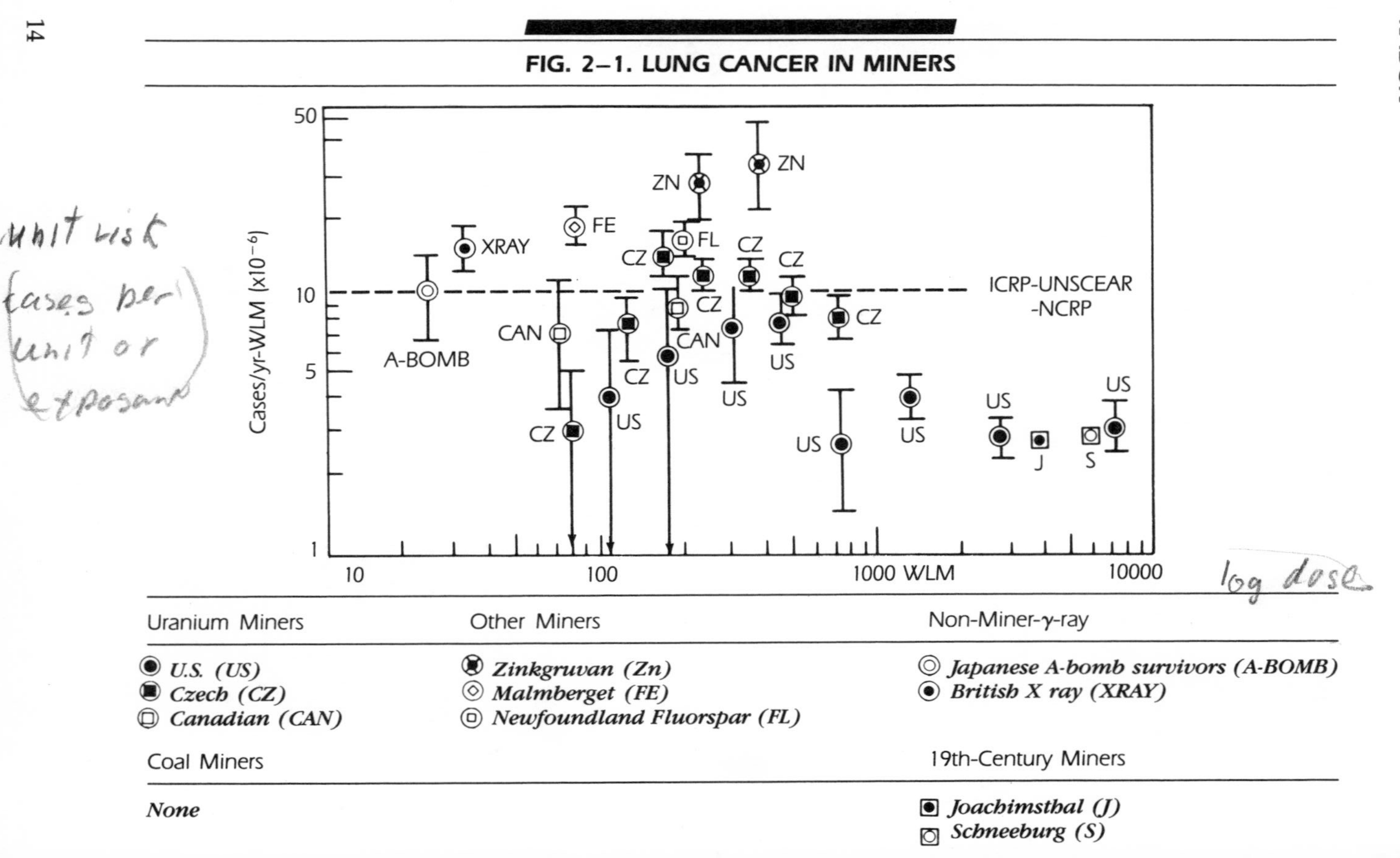

national average. This confirms that radon, and not chemical agents, is very probably the culprit that causes lung cancer in uranium miners.

How Radon Causes Lung Cancer

When radon gas enters the environment, it becomes part of the air, which is actually a mixture of many gases. When a radon atom decays into the other radioactive atoms that we call radon daughters, the latter float around in the air, usually becoming attached to dust particles. When we inhale, these tend to stick to the surfaces of our bronchial tubes. This brings them within about one-thousandth of an inch of some of the most cancer-sensitive cells in the human body, the stem and basal cells of the bronchial epithelium. That is a distance they often can penetrate.

To make matters worse, the alpha particles emitted by radon daughters are about 100 times more efficient than beta and gamma particles at inducing cancer. Part of the reason is that alpha particles have about ten times higher energy. Equally important, they travel ten times more slowly and have twice the electric charge, which means that they exert a much larger force for a much longer time on the molecules they encounter. This allows them to do more extensive damage to a single molecule than can ordinarily be done by beta or gamma particles.

The relevant damage here is to DNA molecules, of which the genes that control the activity of cells are composed. Such damage can have many possible consequences, but in the vast majority of cases it is unimportant. This must be so, since many cases of DNA molecule damage by radiation occur every second in the body of every living person, coming from natural sources of radiation. However, very occasionally, for reasons that are not well understood, this damage to a DNA molecule can cause the cell to divide much more frequently than it normally would. The number of these damaged cells therefore multiplies rapidly and uncontrollably. That is the condition we call cancer.

The damage done by radon daughters sticking to the surfaces of our bronchial tubes is fairly unique. Even most alpha particle emitters, like radium or plutonium (if they could reach the lungs), would do very little harm in a similar situation, as our body has mechanisms for clearing foreign materials out of our bronchial tubes within a few hours. But this does not work for the radon daughters, as they have short half-lives, emitting nearly all their alpha particles within less than an hour.

As a result of this unfortunate combination of circumstances, radon in the air causes radon daughters to administer very high radiation doses to very sensitive cells, and this can result in development of lung cancer ten to fifty or more years later.

The Effects of Cigarette Smoking

One thing that is not yet clearly understood is the role of cigarette smoking in the process we have been describing. Cigarette smoking can not only initiate lung cancers, but it can promote the development of lung cancers initiated by other agents.

Most of our information about lung cancer induction by radon is derived from studies of miners. Among the U.S. and Czechoslovakian miners, the risk of lung cancer from a given radon exposure was last reported as five times greater for smokers than for nonsmokers, but this number has been decreasing as time passes. Some think it is still possible that by the time all of these miners have died, the risk to nonsmokers will have caught up. Among two groups of Swedish miners who have been followed for a much longer time period, the radon risks to smokers and nonsmokers are about equal. Among the Japanese atomic bomb survivors, smokers and nonsmokers seem to run about equal risk of lung cancer resulting from the radiation exposure they received.

All we can say now is that the risk from radon exposure to smokers *may be* somewhat higher than the risk to nonsmokers. At the very least, smokers develop their lung cancers more rapidly and succumb at a younger age.

Picocuries per liter (pCi/l) and Working Levels (WL)

At this point it is necessary for us to introduce the units in which radon concentrations are measured, picocuries per liter (pCi/l). By definition, a curie is the rate of decay of a gram of radium—37 billion decays per second. Because this is a very large quantity, radioactivity in the environment is usually measured in units of one millionth of a millionth of a curie, or a picocurie (pCi).

With a radon concentration of 1 pCi/l, about two alpha particles would be emitted per minute from radon atoms in each liter of air. This is a convenient unit, since the mean radon level in houses throughout the Western world is about 1 pCi/l.

One other unit is sometimes used, the working level (WL), derived from mine regulations. Actually, the working level is a measure of the concentration of radon daughters rather than of radon itself. But as a rough approximation, 1 pCi/l of radon gas is equivalent to 0.005 WL, or 1 WL is equivalent to about 200 pCi/l of radon gas.

Since it is the radon daughters rather than the radon gas that cause lung cancer, one might think that the WL would be the more applicable unit for our consideration. Indeed, that *is* the situation in mines, but *not* in houses. The reason is that it is rather easy to clean the dust out of the air in a house by use of filters or electrostatic precipitators, and this is often done. Since the radon daughters are nearly all attached to dust particles, this reduces the WL drastically. At one time this seemed to be the solution to the radon problem in houses. But we were not that lucky.

When the dust is cleaned out of the air, a radon daughter atom, newly formed from the decay of a radon atom, has fewer dust particles to become attached to. It consequently has a much better chance of remaining unattached or, what is essentially equivalent, of attaching to the extremely tiny dust particles that cannot be removed by the air cleaning device. When we inhale, the larger dust particles to which radon daughters are normally attached have perhaps a 3-percent chance of striking and sticking to the surfaces of the bronchial tubes, where they do their dirty work. But the unattached radon daughter atoms, or those that attach to extremely tiny dust particles when the air is cleaned, move about in all directions with much higher speeds and thus may have a 50-percent or higher chance of striking the surfaces and sticking.

In summary, cleaning the dust out of the air can reduce the number of radon daughter atoms, and hence the WL, rather drastically, but the radon daughter atoms that remain become much more dangerous from the standpoint of the radiation doses they administer. It turns out that these two effects compensate for one another, and the net effect of air cleaning on our radiation exposure is relatively small. There is one possible exception: Cleaning air with ion generators also removes the unattached radon daughters and may therefore provide an important health benefit. This matter is under investigation.

Of course, dust removal does not affect the radon gas, because it doesn't stick to dust particles. Thus the radiation dose is just proportional to the pCi/l of radon gas, irrespective of whether or not the air is cleaned, while the WL changes drastically when the air is cleaned. The pCi/l is therefore the more relevant unit, and we will use it almost exclusively in this book.

Since most of our information on radon effects comes from studies of miners, there is an important question about the differences in radiation exposure, even at 1 pCi/l, between miners and people in their homes. The latter include women and children, whose respiratory systems have rather different dimensions than those of miners. Miners are working, which means that they are breathing more air in and out than people in homes, who are, for the most part, not engaged in strenuous physical activity, but on the other hand, the velocity of the air passing through the miners' bronchial tubes is much higher, which gives the radon daughters less of a chance to stick. The dust conditions in mines are very different from the dust conditions in homes, and that makes a difference.

All of these problems have been considered in great detail by Dr. Naomi Harley and her collaborators from New York University, and their conclusion is that the radiation dose per hour from 1 pCi/l of radon in a home is slightly *larger* than that from 1 pCi/l in a mine. However, the difference is so small that it is often ignored.

The Lung Cancer Risk from Radon Exposure

The horizontal line in Fig. 2–1, which gives a rough estimate of the risk from all of the miner data, corresponds to a risk from lifetime radon exposures in homes of about one death per 10,000 people per year per pCi/l of radon level. This risk persists for about 30 years, which gives a lifetime risk of 30/10,000 or about 1/300. However, this estimate is very oversimplified.

Even when one knows the radiation doses from radon exposure for both groups, it is still not a straightforward task to estimate all radon risks from the data on miners. One problem is that none of the miners were women, elderly men, or children; it is necessary to make assumptions about effects on them. This may be done by making use of data on the Japanese atomic bomb survivors, which included people of all ages and both sexes. Another problem is that most of the miners exposed to high levels of radon in mines are still alive; we can't tell how many of them will eventually die of lung cancer. Because of these problems, different experts have developed varying estimates for the risk of lung cancer from exposure to radon in homes.

There are several ways to describe these estimates, but here we use the number of deaths per year in the United States if, as is widely believed, the average radon level in U.S. homes is 1 pCi/l. The U.S. National Council on Radiation Protection and Measurements (NCRP) estimates this to be 9,000 deaths per year. The National Academy of Sciences Committee on Biological Effects of Ionizing Radiation (BEIR), in its last report (1980), estimates it to be 24,000 deaths per year. The U.S. Environmental Protection Agency (EPA) puts it at 5,000 to 20,000 per year. The Centers for Disease Control (CDC) give it as "up to 30,000." Many other estimates have been published.

For purposes of our discussion we will take this number to be 10,000 deaths per year in the U.S. from radon. If the U.S. population were in age equilibrium, there would be 3 million total deaths per year, so one death in 300 would be due to radon (10,000/3,000,000). This means that living in a house with 1 pCi/l of radon throughout one's life gives one a risk of one chance in 300 of dying from lung cancer due to radon. If your house has 10 pCi/l, your risk is one chance in 30; if it has 100 pCi/l, it is one chance in three.

Note that these estimates are based on the assumption that the risk is *proportional* to the exposure—i.e., ten times the exposure means ten times the risk, one-tenth the exposure means one-tenth the risk, and so on. In scientific circles, this is known as a "linear–no threshold" dose-response relationship. The validity of this assumption has not been tested experimentally, especially at low doses. It is based rather on our understanding of how radiation induces cancer. The basic process here is that a single alpha particle does damage to a DNA molecule. Ten alpha particles will surely damage ten times as many DNA molecules, 100 alpha particles will surely damage 100 times as many DNA molecules, etc. Thus, the number of damaged DNA molecules, and therefore the risk of cancer, is just proportional to the number of alpha particles, which is a measure of the dose.

This conclusion is not inevitable. For one thing, it is well known that there are repair processes. Enzymes run up and down the DNA molecules, repairing the type of damage done by radiation. However, there is no direct evidence that the efficiency of these repair processes depends on the radiation dose. It is therefore assumed that damage has the same probability of being repaired for all doses, which means that the risk remains proportional to the dose.

On the other hand, it seems not unlikely that repair processes will be more efficient if they are not overloaded. This implies that they are more efficient at low doses, and therefore that the risk from low doses is less than implied by the linear–no threshold theory.

This assumption that the cancer risk is proportional to the radiation dose (at least at low doses) is used for all types of radiation, and is responsible for all estimates of the hazards of low-dose radiation. As in the case of alpha particles from radon, this assumption has not been tested experimentally. If it should be found to be invalid, all estimates of the dangers of radiation from nuclear industries, from medical X rays, from radioactive waste, from bomb testing, and from other sources would be greatly reduced. It is generally believed that failure of the simple proportionality (linear) assumption is much more likely for these situations than for the alpha particles from radon.

Another way to express the risk of radon is in terms of the "loss of life expectancy" it causes. To explain this term, consider motor vehicle accidents. If all motor vehicle accident deaths were eliminated, but the probability of all other causes of death at each age remained the same, the average person would live 200 days longer. That does not mean that *each* person would live 200 days longer; the great majority of people who do not die in such accidents would not live any longer than before, while the few who do die in motor vehicle accidents would live many *years* longer. However, averaging over the entire population, the average person would live 200 days longer. We thus say that the loss of life expectancy due to motor vehicle accidents is 200 days.

In terms of loss of life expectancy, each pCi/l of radon in your house (throughout life) reduces your life expectancy by about 25 days. This may be calculated as follows: the average person dying of lung cancer induced by radon loses about 20 years of life. If each person has one chance in 300 of losing 20 years of life, his loss of life expectancy is 1/300 of 20 years, or 25 days. Smoking a pack of cigarettes per day on an average from age 16 reduces life expectancy by about 2,300 days, which is equivalent to the risk of 90 pCi/l of radon in your home ($90 \times 25 = 2{,}300$). Being 25 pounds overweight reduces life expectancy by about 750 days, which is equal to the risk of 30 pCi/l of radon (30 × 25 = 750). If your radon level is 8 pCi/l, this reduces your life expectancy by 200 days (8 × 25), which is equal to your risk from motor vehicle accidents. The 25 days of life expectancy lost from a typical radon level is roughly equal to the risk of dying by drowning, or of dying in a fire. The loss of life expectancy owing to other risks are listed in Table 2–A.

TABLE 2–A. LOSS OF LIFE EXPECTANCY IN DAYS FROM VARIOUS CAUSES

Accidents	male		female
all	669		297
motor vehicle	363		150
pedestrians	49		24
falls	49		52
fire, burns	31		26
drowning	49		11
in home		95	
firearms		11	
poison		17	
suffocation, asphyxiation		13	
reactor meltdown		0.01	
Smoking (males only)			
1–9 cigarettes per day			1,600
20–39 per day			2,500
over 40 per day			3,150
Occupational accidents (by industry)			
all U.S. workers			74
trade			30
manufacturing			45
transportation, utilities			164

construction	300
demolition	1,500
mining	330
fishing	400
forestry	540
Air pollution	12
Explosions	0.4
Radiation	
100 pCi/l of radon in home	2,500
10 pCi/l of radon in home	250
1 pCi/l of radon in home	25
radiation worker in nuclear plant (age 18 to 65)	12
all other natural radiation combined (average citizen)	8
medical X rays (typical)	4
nuclear power (includes reactor accidents, waste management and storage, transport accidents, routine releases or radioactivity, etc.), highest estimate	1.5
nuclear power according to government estimates	0.05
Disease	
heart	2,106
cancer	980
stroke	570
Overweight	
20 lbs.	600
40 lbs.	1,200
Dangerous jobs (20 years of participating)	
auto racers	100
aerialists (high wire)	100
divers	500
boxers	130
Suicide	90
Murder	90

Some people worry about the risk from a chest X ray, but the average American gets as much radiation from radon every three weeks as he gets from a typical chest X ray. Even a radiation worker in a nuclear plant gets only half as much radiation from his job as he typically gets from radon in his home (assuming a level of 1 pCi/l in the latter).

People often ask me whether they should worry about the radon in their home. My answer is, "If you are worried about radiation in any way, shape, or form, you should be very worried about radon." The great majority of Americans get more radiation exposure from radon than from all other sources combined.

Radon in Your Home

There is only one way to find out the extent to which radon in your home is putting you at risk of lung cancer: *Test for it.* There are a number of devices currently available that allow you to conduct a radon test in your home. By following the instructions that accompany the device, and then returning it for evaluation, you can determine your risk. These devices will be discussed and evaluated in a later chapter.

What levels of radon gas constitute unacceptable risk of lung cancer? The Environmental Protection Agency has set the safety standard for radon gas in residences at 4 pCi/l. A person living his whole life in a house with this radon level would have a little over a 1-percent chance (4 chances in 300) of dying from lung cancer due to radon. There is nothing special about 4 pCi/l. It certainly does not mean that 3 pCi/l is "safe" and 5 pCi/l is "unsafe." Obviously the latter is only one and two-thirds times as risky as the former. Other agencies have adopted other recommendations. The U.S. National Council on Radiation Protection and Measurements (NCRP) uses 8 pCi/l, and the Swedish government uses 20 pCi/l.

I believe that the answer to "what level is acceptable" is that every person must decide for himself what risk is acceptable.

The information presented in the above explanation of the risk can tell you ***how big*** your risk is, if you know the radon level in your home. Then you have to decide what risk ***you*** are willing to accept.

Recent History of the Radon Problem

The risk estimates we have been discussing are not appreciably different from those derived 15 years ago. But the public has only recently begun to learn what the scientific community has known for over a decade.

Radon exposure to the public first became newsworthy in connection with uranium mill tailings. Uranium is mined to produce fuel for nuclear reactors. Since even the richest ore contains only a very small percentage of uranium, transportation costs are minimized by building a mill very close to the mine for doing chemical separation of the uranium. Everything but the uranium ends up as waste in huge piles called *tailings*, near the mill. The tailings, of course, include the radium that is always associated with uranium, and these tailings piles therefore emit radon in great abundance.

In 1972, some EPA scientists calculated that radon emitted from these piles is, by far, the most serious source of public radiation exposure from the nuclear power industry, thousands of times more harmful in the long run than radioactive waste produced by nuclear power plants. This led to lots of government research, and it was soon found that covering these tailings piles with dirt greatly reduced the radon emissions; this is now required by law.

Radon next came to prominence in the news when it was found that some of these mill tailings had been used for construction of houses and schools, especially in Grand Junction, Colorado. This caused high radon levels in those structures. These houses had been lived in for about 25 years. An epidemiology study was done and found no excess lung cancer for those who lived in the houses. Nevertheless, the high radon levels led to a large and expensive government program to dig this material out.

An interesting aspect of this work was that in measuring radon levels in houses built with these mill tailings, the government also made measurements in several houses where mill tailings had not been used, and found that radon levels in them were often not very much lower. This was one of the first indications to the scientific community that radon levels in normal houses were quite high.

A similar situation developed in connection with phosphate mining in Florida. Phosphate rock contains a lot of uranium, and phosphate lands therefore emit a lot of radon. When this problem was studied, it was found that homes built on phosphate lands had higher radon levels than nearby homes not built on phosphate lands, although the difference was not dramatic. This led to government-sponsored research to find the cause of high radon levels in normal houses. Several suspicious sources were investigated—concrete, plaster, natural gas, water supplies, etc. All of these made small contributions, but it was finally concluded that the predominant source of radon in houses was simply the ground on which they were built. It was a purely natural phenomenon, with no one to blame.

At this point, however, U.S. government-sponsored research on the problem all but stopped. Official Environmental Protection Agency policy was that natural sources of radiation were not the government's business. There was nothing new in this policy. Many billions of dollars per year have been spent in minimizing the public's exposure to radiation from the nuclear industry, although it has always been recognized that this is far less than 1 percent of its exposure to natural radiation, about which nothing is being done.

An interesting example of this policy is a uranium mill tailings pipe in Salt Lake City, Utah. Mining and milling had ceased long before the city grew to surround the pile. When concern arose about the radon being emitted from it, the U.S. government took responsibility and is now moving the pile out into the desert, an extremely expensive operation involving over 200 railroad carloads per day for over two years. Before undertaking this project, no one bothered to study the radon levels in nearby homes. It is quite probable that they are predominately due to radon naturally seeping out of the ground rather than to that which is transported through the air from the tailings pile.

While U.S. government policy was to ignore the problem of radon in homes, other nations showed more concern. Large-scale studies were carried out in almost every country in Western Europe. There were ten times as many measurements in Canada as in the United States, although the Canadian population is only one-tenth as large. The last two U.S. universities with federal research funding for research on radon in homes had their grants terminated early in 1985—the Radon Project at the University of Pittsburgh was one of these.

But then a truly dramatic event changed all this. A new nuclear power plant was preparing to start up near Pottstown, Pennsylvania, and its management began requiring workers to pass through a radiation monitoring portal to check whether any radioactive contamination was being carried out as they left the plant. Curiously, one of the workers, Stanley Watras, set off the alarm on *entering* the plant. The problem was traced to radon in his home. In fact, to this day his house still holds the world record, 2,700 pCi/l. At this point the Pennsylvania Department of Environmental Resources undertook a program of measuring radon levels in nearby houses, and many of them were also found to have very high radon levels. This led to lots of media publicity, which led to more measurements, which led to discovery of more houses with high radon levels, which led in turn to more widespread publicity, and so on. When the public becomes concerned, politicians respond, and government agencies begin to move. Radon was becoming a household word.

3

The Geology Connection

As noted, when the 2,700 pCi/l radon level in the Watras house was discovered, the Pennsylvania Department of Environmental Resources began studying other houses in the neighborhood. The house next door had a level lower than 1 pCi/l!. Several other nearby houses had levels of many hundreds of pCi/l, but most of them had normal levels. How can this be?

There are two geological features necessary to feed large amounts of radon into a house: a rich deposit of uranium somewhere in the ground, and an easy pathway for radon from that deposit to the site of the house. Through normally packed soil, radon can diffuse only a few feet in its limited lifetime. Half of all radon atoms decay within 3.8 days, half of the rest decay within the next 3.8 days, and so forth. However, if there is a crack in the rock, radon can percolate up to the surface from depths of hundreds of feet. If the soil is loose, dry, and grainy, radon gets through much more easily than, for example, if it is a wet clay. Unfortunately, cracks in the rock and porous paths through soil are not easy to recognize or to find, even with fairly elaborate investigation. Whether or not there is such a pathway leading to a given house is strictly a matter of luck.

The other geological ingredient in causing radon problems, a rich uranium deposit, is also highly unpredictable. The deposit under the Watras house was only several feet in extent, and this is not atypical. Uranium in rock is easily dissolved under certain chemical conditions—*acid* and *oxidizing*, for those who know chemistry. When these conditions occurred, many millions of years ago, the uranium from rock over a large area was dissolved and it moved with the groundwater until it came to a place where the chemical conditions changed (to *alkaline* and *reducing*). There, all of the uranium precipitated out of solution and became a surface coating on the grains of the local rock, forming a rich uranium deposit. Incidentally, the fact that the uranium is a thin coating on the surface of rock grains explains why the radon atoms formed from it have such a good chance of getting into the pore spaces between the grains, allowing them to percolate up to the surface. If the uranium were mixed through the volume of the grains, the radon atoms would rarely escape from the grain interiors.

Thus, the locations of rich uranium deposits are determined by where the chemical conditions in the groundwater changed many millions of years ago. It is all but impossible to predict these locations, or to tell whether your house is above one of them.

For reasons that go even further back into geological history, some types of rocks have more uranium than others. On the average, rock contains 2.7 parts per million (ppm) of uranium; that is, there are 2.7 pounds of uranium in every million pounds of rock. But granite contains an average of 4.7 ppm, and shale averages 3.7 ppm, whereas sandstone and basalt average only about 0.5 ppm and 0.9 ppm respectively. Areas where the rock is granite are therefore more likely to have radon problems than where the rock is sandstone. But that doesn't mean that there aren't high radon areas built on sandstone. Geology is highly variable and unpredictable.

Some granite formations, like the Conway granite of New Hampshire, have much higher average uranium content than others, and the same is true of shales—the Chattanooga shale that underlies areas of Tennessee, Kentucky, and Ohio has much more uranium than most shales. In fact, some foresee the day when the Chattanooga shale and the Conway granite will be mined for their uranium content.

There are large variations in uranium content within a given formation. For example, the Chattanooga shale has much higher uranium content in parts of Tennessee than in Ohio. In most cases, these variations have never been studied. But even where they have been studied, they are poor predictors of radon problems. The Watras house is on the Reading Prong, a granite formation that extends from near Reading in southeastern Pennsylvania, through a wide band of northern New Jersey (e.g., Morris County), a narrow band in New York State (e.g., Putnam County), and into Connecticut. This whole formation is known to have high uranium content and hence was expected to have radon problems. However, measurements now make it clear that the radon problems are largely confined to the Pennsylvania section plus extreme western New Jersey.

In summary, geologists have had great difficulty in predicting what areas might have radon problems. In 1985, maps of where to expect high radon levels appeared in many places, all based on geological evidence. Since granites and certain types of shale have higher than average uranium concentrations, areas underlain with these rocks were designated as having the potential for radon problems. Actually, uranium concentrations in these rocks average only about twice those in other rocks. Moreover, the porosity of the overlying soil and cracks in the rocks were not taken into account in preparing the maps; in fact, very little is known about these factors. When large-scale radon measurements began to become available in mid-1986, it became clear that these maps were useless.

In fact, they were worse than useless, because they caused undue alarm in some areas and undue complacency in others. For example, Morris County in northern New Jersey and Camden County in southern New Jersey have about equal populations, but owing to geological predictions, hundreds of times as many radon measurements have been made in Morris as in Camden. The results, however, indicate roughly equal radon levels in the two. As another example, the Pennsylvania Department of Environmental Resources, spurred on by geological evidence, made more than 20,000 measurements in the three counties traversed by the Reading Prong, completely ignoring the rest of the state. It is now clear that several other Pennsylvania counties have comparable or higher radon levels than these.

As a result of the heavy publicity, a substantial fraction of all interest in radon has been concentrated on the Reading Prong, but the publicity has spilled over to neighboring areas, leading to a reasonable number of measurements nearby. The two highest radon levels I have encountered in my studies, 2,500 pCi/l and 1,200 pCi/l, were in houses near to, but *not on*, the Reading Prong. Because radon cannot diffuse more than a few yards sidewise in the ground, the high radon levels in these houses could not be attributed to the Prong.

Distribution of Radon Levels Within a County

Although they are highly unpredictable, geological factors do cause median radon levels to vary substantially from one county to another. But within a county, the variations from house to house are even larger. In a county where the median radon level is 1 pCi/l (that is, half of all houses have levels above 1 pCi/l, and the other half have levels below that figure), typically 15 percent of all houses will be above 3 pCi/l, 2 percent of all houses will be above 10 pCi/l, and up to 1 percent will be above 20 pCi/l. On the other hand, if the median level in a county is 3 pCi/l, 15 percent of all houses will be below 1 pCi/l.

Most people would call a county with a median level of 3 pCi/l a high-radon area, and a county with a 1 pCi/l median a low-radon area. But from these examples we see that 15 percent of the houses in the low-radon area have higher levels than *most* houses in the high-radon area. And 15 percent of the houses in the high-radon area have lower radon levels than *most* houses in the low-radon area. Clearly, the median radon level in your county is far from a conclusive indicator of the radon situation in your home.

That is not to say it is not a useful indicator. It is a much better indicator than anything obtainable from geological evidence. If your degree of concern is such that you would take remedial action if the radon level in your house was above 10 pCi/l, the probability for your house to be above that action level is 2 percent if the median level in your county is 1 pCi/l, but it is about 15 percent, or roughly seven times higher, if the median level in your county is 3 pCi/l. Data on median radon levels in counties would therefore be very useful.

Mean Radon Levels in Various Counties

Unfortunately, data on median radon levels in counties are not generally available because studies are still at a very early stage. A scientific study requires measurements in a random selection of houses, but owing to lack of financial resources, very few such studies have been done. With a few exceptions to be mentioned below, all data are from measurements purchased by householders. This severely limits the number of measurements in lower-income households, especially where the price of a measurement is $50 or more, as has often been the case. A person is more likely to purchase a measurement if a nearby house has a high radon level, and a house with a high level is more likely to get further measurements; both of these facts bias averages derived from the data, making them higher than unbiased (random) measurements would show.

Fig. 3–1 shows a comparison between results of purchased and random measurements in the same counties. There are lots of other biases—factors that can make the data unrepresentative

FIG. 3–1. COMPARISON BETWEEN PURCHASED MEASUREMENTS (PM) AND RANDOM-SELECTION, NO-CHARGE (RS-NC) MEASUREMENTS

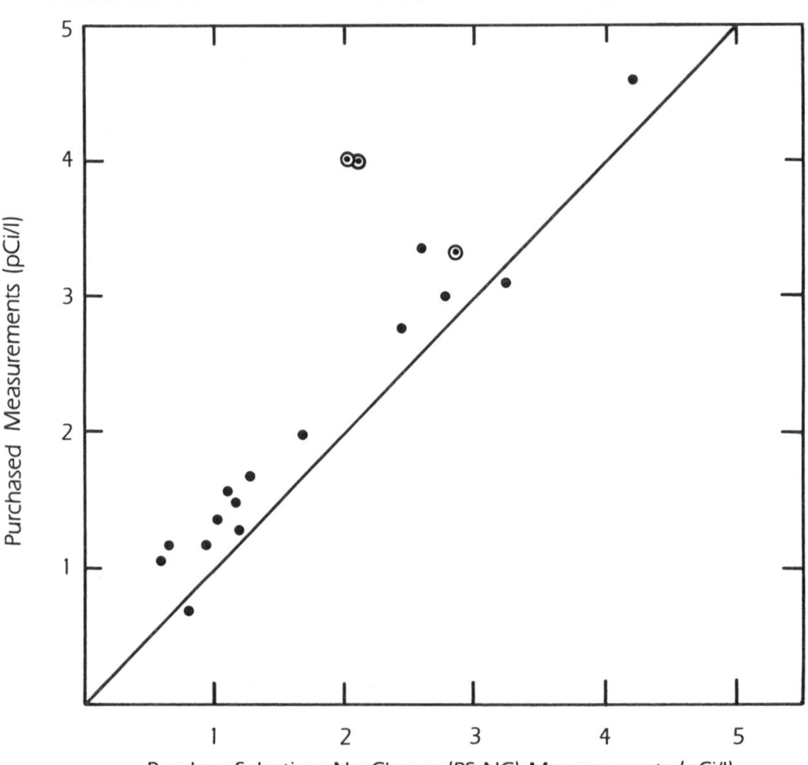

Each point represents a certain county; its horizontal position is the mean radon level obtained in a RS-NC study, and its vertical position is the mean radon level obtained from purchased measurements. If the two studies gave identical results, all points would be on the diagonal line. The fact that they lie above the diagonal line means that purchased measurements give higher mean values, as might be expected from the biases discussed above.

The three points surrounded by circles are for the Reading Prong counties of Pennsylvania. In those countries, most purchased measurements were probably from people who lived on the Reading Prong geological formation, the exact location of which was widely advertised; on the other hand, the RS-NC study chose houses from throughout those countries, the great majority of which were not *on the Reading Prong. This explains why those points are far above the diagonal line.*

of real risk—in most compilations. Radon levels in houses are typically 60 percent higher in winter than in summer, but the great majority of measurements have been in winter. Some people keep windows that would normally be open closed during the measurement period. A substantial fraction of all measurements have been made in basements, where radon levels are generally two or three times higher than in the main living areas of the house (because radon enters through the basement). Average levels in the lived-in portions of homes would certainly be lower.

In our program at the University of Pittsburgh, we have made an effort to understand and reduce these biases by accompanying all measurements with a questionnaire and by doing several thousand random-selection, no-charge measurements. For example, we eliminate houses in which radon levels have been measured previously, and we separate measurements in basements from those in the living areas of the house. In addition, we have developed empirically derived corrections for seasonal variations, and for abnormal window closing during the test, and these are applied to the data. Our tests costs $12, which makes them widely affordable. On our questionnaires we ask about socioeconomic status and have found that radon levels in houses are not correlated with the value of the house (where it is above $40,000), with the annual household income (where it is above $15,000), or with the educational attainment of the head of the household. Houses valued at less than $40,000 and in which household income is below $15,000 per year so far seem to have about 25 percent lower radon levels than others, but this matter is being further investigated by offering free measurements to low-income families.

Probably the most important source of bias in our data is that measurements are much more frequently purchased in areas where there has been lots of publicity about radon, which normally is in areas where high radon levels have been reported or are suspected. Median radon levels in our "selected" and corrected data are still somewhat higher than in our random-selection, no-charge studies, as is evident from Fig. 3–1.

In the tables that follow, we present the University of Pittsburgh data obtained up to January 15, 1987, for states, for counties, and for zip codes in which we have at least four measurements in the main living areas (not basements). We list the geometric mean radon level, which is very close to the median, and the average radon level, which gives added weight to high measured values. We also list the number of measurements as an indication of the uncertainty in the mean radon level. Where there are only five measurements, the uncertainty owing to small sample size is about 50 percent, for 20 measurements it is perhaps 25 percent, and for 100 measurements it is something like 12 percent. These uncertainties are in addition to the uncertainty resulting from bias in the whole data set. Also listed in the tables are the number of measurements between 4 and 20 pCi/l, and over 20 pCi/l.

These tables may be useful for understanding broadly how radon levels vary with geography. They give an indication of the *probability* for a high radon level to be found in a house, but do not give an estimate of that level.

Correlations with House Characteristics

In past and present studies, many questionnaire items have dealt with house characteristics to determine how they might affect radon levels. We review some of our findings here.

One very important question is the effect of the recent emphasis on weatherizing houses to reduce energy costs for heating and cooling. This includes weatherstripping, sealing around windowpanes, closing gaps under doors to the outside, installing storm windows, and so forth. Since these actions reduce air exchange with the outdoors, they trap radon indoors longer and hence tend to increase the radon concentration inside the house. Several cases have been reported where super-insulated houses have had high radon levels. But for the great majority of cases, these measures, carried out since the "energy crisis" of 1974, have increased radon levels by about 10 percent in houses where they were employed. On the other hand, houses that are generally tight and well weatherized average about 40 percent higher radon levels than poorly weatherized or drafty houses.

Since more attention has been paid to weatherization in constructing houses in recent years, one might think that new houses would have higher average radon levels, but of course better sealing may also help to prevent radon from entering the house. Our surveys indicate that mean radon levels do tend to decrease as the age of the house increases, but the effect is not large. Sixty-year-old houses average about 25 percent lower radon levels than new houses, but for houses over 80 years old, the decrease with age seems to reverse.

Houses in especially windy locations have about 15 percent higher radon levels than average because wind increases the vacuum inside the house, sucking more radon out of the ground. On average, rural houses have radon levels about 20 percent higher than urban houses, with suburban houses somewhere in between.

Construction materials—wood, stone, or brick—do not affect radon levels. Houses with basements average about 30 percent higher radon levels than houses without basements, because the former have more contact area with the ground through which radon can enter. At one time it was thought that houses built on a concrete slab could not have radon problems, but this difference of only 30 percent applies to them specifically, and the second highest radon level we have found—1,200 pCi/l—was in a house built on a concrete slab.

Houses with forced-air heating systems have about 20 percent higher radon levels in the living areas than those with steam or hot-water heating systems, probably because the former circulate more air from the basement. There is little consistent difference in radon levels among living rooms, dining rooms, kitchens, bedrooms, and halls. Family rooms average about 40 percent higher because they are often in the basement. No difference was found between kitchens that use gas and electricity for cooking.

Several other features of houses have been explored, but none of these seems to have an important effect. In fact, it is surprising how little effect all house characteristics have on radon levels. Clearly, the location of the house is much more important than its characteristics in determining the extent of its radon problems.

This reinforces the basic conclusion of this chapter. You can't depend on any information that is available to estimate the radon level in your home. The only way to know whether you have a radon problem is to get a measurement.

TABLE 3–A. RADON LEVELS (pCi/l) BY STATE AND COUNTY

The left side of table is for living areas of the house and the right side is for basements and cellars. Houses in which there has been a previous measurement are excluded. The first column is the number of measurements, the second and third are the mean and average values of the radon levels for these measurements in pCi/l, and the fourth and fifth are the number of these measurements between 4–20 pCi/l and over 20 pCi/l, respectively. Since our computer entries include only the first three letters in the name of a county, data for counties with the same first three letters are combined. Where one of these has a much larger population than the others, it is listed with an asterisk (*). Data obtained from random-selection, no-charge measurements are designated by a dagger (†). All other data are from purchased measurements.

	Living Areas					Basements				
State/County	NMBR	Mean	Aver.	4–20	>20	NMBR	Mean	Aver.	4–20	>20
Alabama	71	1.93	3.30	11	1	8	7.20	12.01	2	3
Cullman	5	1.20	1.60	0	0	–	–	–	–	–
Jefferson	13	1.63	2.43	1	0	3	1.80	1.80	0	0
Madison	17	4.01	7.00	7	1	–	–	–	–	–
Shelby	4	2.30	2.30	0	0	–	–	–	–	–
Alaska	35	1.00	1.50	4	0	–	–	–	–	–
Anchorage	11	0.75	1.10	0	0	–	–	–	–	–
Haines	4	1.15	1.20	0	0	–	–	–	–	–
Arizona	101	1.83	2.60	13	1	4	6.60	6.80	4	0
Cochise*	16	1.55	2.60	3	0	–	–	–	–	–
Maricopa	42	1.81	2.70	2	1	–	–	–	–	–
Pima	15	1.60	1.90	1	0	–	–	–	–	–
Yavapai	11	1.80	2.30	2	0	–	–	–	–	–
Arkansas	53	1.51	2.20	7	0	18	2.30	5.53	2	2
Baxter	5	1.20	1.32	0	0	–	–	–	–	–
Marion	4	2.05	2.71	2	0	–	–	–	–	–
Pulaski	7	1.40	1.52	0	0	–	–	–	–	–
Sebastien†	34	0.62	0.80	0	0	–	–	–	–	–

State/County	Living Areas					Basements				
	NMBR	Mean	Aver.	4–20	>20	NMBR	Mean	Aver.	4–20	>20
Washington	9	1.15	2.80	1	0	–	–	–	–	–
California	355	1.00	1.50	14	1	10	2.15	3.61	4	0
Alameda	19	0.80	1.12	1	0	–	–	–	–	–
Contra Costa	5	0.83	1.15	0	0	–	–	–	–	–
Contra Costa†	42	0.80	1.01	0	0	–	–	–	–	–
Humboldt	4	0.60	0.72	0	0	–	–	–	–	–
Humboldt†	54	0.90	1.30	3	0	–	–	–	–	–
Lassen	6	2.00	2.52	2	0	–	–	–	–	–
Los Angeles	74	1.00	1.31	3	0	–	–	–	–	–
Marin	13	0.82	1.00	0	0	–	–	–	–	–
Monterey	4	0.60	0.70	0	0	–	–	–	–	–
Nevada	4	0.74	0.83	0	0	–	–	–	–	–
Orange	20	1.03	1.20	0	0	–	–	–	–	–
Placer	8	1.70	3.40	1	0	–	–	–	–	–
Riverside	16	0.90	1.12	0	0	–	–	–	–	–
San ———	83	0.93	1.30	4	0	–	–	–	–	–
San Mateo†	16	0.70	0.93	1	0	–	–	–	–	–
Solano	4	0.93	0.95	0	0	–	–	–	–	–
Solano†	18	0.82	1.00	0	0	–	–	–	–	–
Sonoma	16	0.82	1.00	0	0	–	–	–	–	–
Ventura	7	1.73	1.94	0	0	–	–	–	–	–
Ventura†	30	1.43	1.72	2	0	–	–	–	–	–
Colorado	456	3.40	5.35	166	16	123	5.75	9.65	61	16
Arapahoe	24	3.42	4.20	12	0	8	7.85	8.60	7	0
Boulder	78	3.30	4.85	25	3	26	4.74	6.05	15	2
Boulder†	50	2.60	3.71	15	0	–	–	–	–	–
Chaffee	5	3.71	4.10	3	0	–	–	–	–	–
Clear Creek	4	2.34	2.80	1	0	–	–	–	–	–
Denver	16	3.54	4.63	9	0	5	3.98	4.71	2	0
Douglas	6	2.41	2.90	1	0	3	6.25	8.20	2	0

El Paso	143	3.30	5.20	44	7	44	5.70	10.64	14	10
Jefferson	90	3.95	7.64	36	5	11	8.33	17.11	7	1
Jefferson†	6	5.10	5.42	5	0	—	—	—	—	—
Larimer	21	3.14	4.40	8	0	18	7.50	12.35	11	3
Mesa	7	1.51	1.70	0	0	—	—	—	—	—
Montrose*	4	3.93	4.43	2	0	3	3.94	4.60	2	0
Pueblo	4	3.15	3.53	1	0	—	—	—	—	—
Teller	16	4.85	5.71	10	0	—	—	—	—	—
Weld	6	1.95	2.34	1	0	—	—	—	—	—
Connecticut	551	1.20	1.95	53	1	149	2.80	6.01	44	8
Fairfield	210	1.15	2.04	21	1	43	3.51	6.60	15	4
Hartford	80	1.00	1.43	3	0	38	2.41	8.50	10	2
Litchfield	55	1.50	2.30	6	0	16	2.91	4.74	3	1
Middlesex	32	1.34	2.10	3	0	10	1.73	2.70	3	0
New Haven*	102	1.20	1.90	8	0	21	3.10	4.22	8	0
Toland	33	1.40	2.10	4	0	9	1.80	3.10	1	0
Windham	6	2.90	3.80	3	0	—	—	—	—	—
D.C.	159	0.91	1.23	7	0	49	1.62	2.40	10	0
Delaware	24	0.80	0.90	0	0	6	4.34	5.45	4	0
Delaware†	62	0.63	0.83	2	0	—	—	—	—	—
Kent†	10	0.52	0.53	0	0	—	—	—	—	—
New Castle†	33	0.70	0.90	1	0	5	3.80	4.80	3	0
Sussex	7	0.52	0.52	0	0	—	—	—	—	—
Sussex†	17	0.60	0.65	0	0	—	—	—	—	—
Florida	459	1.30	2.40	49	4	4	0.80	0.90	0	0
Alachura	4	2.30	3.20	1	0	—	—	—	—	—
Brevard	4	0.65	0.71	0	0	—	—	—	—	—
Broward	15	0.85	1.04	0	0	—	—	—	—	—
Charlotte	4	1.60	3.21	1	0	—	—	—	—	—
Citrus	12	2.41	3.50	2	0	—	—	—	—	—
Clay	6	0.85	1.10	0	0	—	—	—	—	—
Dade	7	1.60	2.10	1	0	—	—	—	—	—

State/County	Living Areas					Basements				
	NMBR	Mean	Aver.	4–20	>20	NMBR	Mean	Aver.	4–20	>20
Duval	47	0.74	0.85	1	0	–	–	–	–	–
Duval†	13	0.60	0.70	0	0	–	–	–	–	–
Hernando	7	1.70	2.92	2	0	–	–	–	–	–
Hillsborough	80	1.33	1.83	6	0	–	–	–	–	–
Lake	5	1.30	1.50	0	0	–	–	–	–	–
Leon	5	1.70	3.20	1	0	–	–	–	–	–
Manatee	11	1.04	1.31	0	0	–	–	–	–	–
Marion*	14	1.50	2.80	2	0	–	–	–	–	–
Okaloosa	13	0.91	1.00	0	0	–	–	–	–	–
Orange	4	0.80	1.43	1	0	–	–	–	–	–
Palm Beach	12	0.80	1.30	1	0	–	–	–	–	–
Pasco	7	0.63	0.70	0	0	–	–	–	–	–
Pinellas	26	1.20	2.70	1	1	–	–	–	–	–
Polk	78	2.51	5.30	23	2	–	–	–	–	–
Sarasota	45	1.12	1.60	3	0	–	–	–	–	–
Sarasota†	59	0.90	1.20	2	0	–	–	–	–	–
Seminole	4	0.50	0.50	0	0	–	–	–	–	–
St. Lucie	5	1.40	2.82	1	0	–	–	–	–	–
Volusia	6	0.65	0.80	0	0	–	–	–	–	–
Georgia	151	1.63	2.20	15	0	24	3.40	4.70	10	1
Chatham*	12	0.82	1.10	0	0	–	–	–	–	–
Chatham†	29	0.85	1.24	1	0	–	–	–	–	–
Cherokee	6	1.95	4.30	2	0	–	–	–	–	–
Clayton*	4	2.91	3.40	2	0	–	–	–	–	–
Cobb	16	1.51	1.71	0	0	4	4.13	4.43	2	0
Coweta	4	0.81	1.20	0	0	–	–	–	–	–
DeKalb	23	2.02	2.30	2	0	4	3.00	3.10	1	0
Fulton	25	1.94	2.65	4	0	7	3.54	3.90	3	0
Glynn	4	0.52	0.60	0	0	–	–	–	–	–
Gwinett	9	2.30	2.75	1	0	–	–	–	–	–

Richmond	5	1.05	1.22	0	0	–	–	–	–	–
Idaho	832	2.10	3.60	167	9	60	3.03	3.80	26	0
Ada	147	2.12	2.70	30	0	18	3.15	3.81	7	0
Bannock	20	1.70	2.24	4	0	–	–	–	–	–
Benewah	9	1.54	2.33	1	0	–	–	–	–	–
Blaine	51	4.42	6.83	28	1	3	4.14	4.55	2	0
Boise	23	2.10	2.60	3	0	–	–	–	–	–
Bonneville*	53	1.70	2.12	2	0	4	2.03	2.70	1	0
Boundary	14	1.70	3.22	1	1	–	–	–	–	–
Canyon	59	1.40	1.73	1	0	5	1.62	1.90	0	0
Caribou	8	1.32	1.70	1	0	–	–	–	–	–
Cassia	8	1.20	2.30	1	0	–	–	–	–	–
Clearwater	15	1.70	2.13	3	0	–	–	–	–	–
Custer	4	3.20	5.92	1	0	–	–	–	–	–
Elmore	18	4.14	7.63	8	1	–	–	–	–	–
Franklin	5	0.73	0.74	0	0	–	–	–	–	–
Fremont	9	2.63	2.74	0	0	–	–	–	–	–
Gem	12	2.05	2.12	0	0	–	–	–	–	–
Gooding	4	1.92	2.05	0	0	–	–	–	–	–
Idaho	26	1.74	2.90	6	0	–	–	–	–	–
Jerome	9	1.51	2.00	1	0	–	–	–	–	–
Kootenai	94	4.00	8.83	38	5	6	4.40	5.60	4	0
Latah	27	1.51	2.20	4	0	–	–	–	–	–
Latah†	32	1.10	2.20	3	0	–	–	–	–	–
Lemhi	6	2.40	5.40	3	0	–	–	–	–	–
Lewis	9	1.53	1.90	1	0	–	–	–	–	–
Lincoln	4	3.00	4.13	2	0	–	–	–	–	–
Minidoka	7	3.00	3.43	1	0	–	–	–	–	–
Nez Perce	41	1.93	2.94	4	1	–	–	–	–	–
Owyhee	15	1.80	2.42	3	0	–	–	–	–	–
Payette	13	1.92	2.45	2	0	–	–	–	–	–
Shoshone	13	2.51	4.20	5	0	–	–	–	–	–

State/County	Living Areas					Basements				
	NMBR	Mean	Aver.	4–20	>20	NMBR	Mean	Aver.	4–20	>20
Teton	4	10.00	10.45	4	0	–	–	–	–	–
Twin Falls	15	2.04	2.60	3	0	–	–	–	–	–
Valley	43	1.02	1.34	2	0	6	2.90	3.83	4	0
Washington	12	1.82	1.90	0	0	–	–	–	–	–
Illinois	1650	1.61	2.44	244	7	663	2.61	4.00	186	11
Champaign	6	1.95	2.90	1	0	4	3.84	4.14	2	0
Cook	873	1.40	2.13	95	4	313	2.05	3.12	66	2
DeKalb	11	1.50	1.80	0	0	6	3.02	3.80	2	0
DuPage	292	2.03	3.05	66	1	172	3.15	4.55	61	2
Grundy	6	1.21	1.31	0	0	–	–	–	–	–
Iroquois	5	1.70	1.90	0	0	–	–	–	–	–
Jo Daviess	5	4.24	4.94	2	0	–	–	–	–	–
Kane*	52	2.30	3.10	14	0	23	4.43	6.91	8	3
Kendall	7	2.60	2.80	1	0	3	3.50	4.74	1	0
Lake	144	1.40	1.81	13	0	53	2.85	4.05	11	2
LaSalle	8	1.91	3.93	1	0	3	3.20	4.80	1	0
Logan	4	3.34	4.50	2	0	–	–	–	–	–
Macon*	8	4.10	4.90	4	0	–	–	–	–	–
Madison	8	1.01	1.10	0	0	–	–	–	–	–
McHenry	31	1.90	2.61	4	0	9	4.10	7.00	3	1
McLean	5	3.90	6.90	2	0	3	3.70	4.00	2	0
Morgan	6	2.93	6.00	2	1	–	–	–	–	–
Peoria	5	3.92	6.60	3	0	–	–	–	–	–
Rock Island	6	1.60	2.24	1	0	–	–	–	–	–
Sangamon	8	3.31	4.80	4	0	3	5.00	5.03	3	0
St. Clair	8	2.34	2.54	1	0	–	–	–	–	–
Tazewell	8	4.41	7.26	3	1	–	–	–	–	–
Vermillion	6	1.14	1.21	0	0	–	–	–	–	–
Whiteside*	5	1.90	2.00	0	0	–	–	–	–	–
Will*	55	1.92	2.70	13	0	21	4.00	5.30	9	0

Winnebago	4	1.61	2.00	1	0	–	–	–	–	–
Indiana	280	2.40	4.90	61	13	108	4.00	7.20	43	8
Allen	5	3.43	10.70	2	1	–	–	–	–	–
Carroll	8	4.65	14.73	2	2	9	16.80	23.00	5	3
Clark*	64	2.37	5.00	14	3	37	5.12	8.30	16	3
Delaware	4	4.52	5.00	2	0	–	–	–	–	–
Floyd	5	2.00	2.25	0	0	–	–	–	–	–
Lake	27	1.40	2.60	1	1	6	1.50	2.00	1	0
LaPorte	9	2.10	2.85	2	0	–	–	–	–	–
Madison	8	4.14	7.50	2	1	–	–	–	–	–
Madison†	36	2.00	2.73	5	0	–	–	–	–	–
Marion	30	3.80	5.34	13	1	7	6.90	11.92	3	2
Monroe*	5	1.80	2.21	1	0	7	4.20	4.70	4	0
Porter	10	1.92	2.80	1	0	–	–	–	–	–
Putnam	4	3.64	4.54	2	0	–	–	–	–	–
Tippecanoe*	6	2.40	2.90	1	0	3	4.10	4.30	2	0
Vigo	25	2.50	3.10	4	0	–	–	–	–	–
Iowa	78	3.01	4.70	30	2	20	4.20	5.61	11	0
Black Hawk	4	2.80	3.31	1	0	3	7.60	7.83	3	0
Dickinson	5	1.83	2.44	1	0	–	–	–	–	–
Johnson	4	2.45	2.74	1	0	–	–	–	–	–
Linn†	52	1.21	1.62	4	0	–	–	–	–	–
Polk	5	3.22	5.10	2	0	3	8.20	8.24	3	0
Scott	7	3.30	4.13	3	0	–	–	–	–	–
Story	4	3.71	5.50	2	0	–	–	–	–	–
Winneshak*	4	2.60	3.43	1	0	–	–	–	–	–
Kansas	93	2.50	3.85	27	1	32	3.52	5.50	13	2
Douglas	4	3.41	5.22	2	0	–	–	–	–	–
Johnson	52	2.80	4.40	16	1	23	3.00	4.61	9	1
Leavenworth	4	0.94	1.00	0	0	–	–	–	–	–
Kentucky	127	2.60	4.33	44	2	24	4.31	7.35	8	2
Boone	4	1.20	1.30	0	0	–	–	–	–	–

State/County	Living Areas					Basements				
	NMBR	Mean	Aver.	4–20	>20	NMBR	Mean	Aver.	4–20	>20
Boyd*	4	3.30	6.25	2	0	–	–	–	–	–
Campbell†	19	0.72	0.94	0	0	–	–	–	–	–
Fayette	26	4.70	7.01	15	1	3	7.82	11.00	1	1
Fayette†	35	2.80	6.94	13	1	–	–	–	–	–
Hardin*	6	1.54	2.44	1	0	–	–	–	–	–
Jefferson	33	2.62	3.80	10	0	11	4.35	7.20	5	0
Kenton	4	3.73	6.84	1	0	–	–	–	–	–
Madison	8	1.85	2.82	2	0	–	–	–	–	–
Louisiana	26	0.90	1.12	1	0	–	–	–	–	–
Jefferson*	5	0.60	0.70	0	0	–	–	–	–	–
Maine	291	1.40	2.51	34	3	35	3.41	6.20	13	1
Androscoggin	28	1.50	2.34	4	0	–	–	–	–	–
Aroostook	8	1.00	1.10	0	0	–	–	–	–	–
Cumberland	57	2.43	4.10	14	1	6	12.83	17.40	4	1
Franklin	9	1.10	1.62	1	0	–	–	–	–	–
Hancock	35	1.20	3.33	4	1	5	3.60	5.44	4	0
Kennebec	11	2.40	4.84	0	1	3	1.50	2.31	0	0
Knox	17	1.00	1.33	0	0	3	3.90	4.44	1	0
Lincoln	25	1.00	1.70	3	0	–	–	–	–	–
Oxford	15	2.34	3.23	5	0	–	–	–	–	–
Penobscot	22	0.80	1.14	1	0	4	2.64	3.00	1	0
Sagadaboc	12	1.10	1.82	1	0	–	–	–	–	–
Somerset	4	1.10	1.40	0	0	–	–	–	–	–
Waldo	13	1.10	1.30	0	0	4	1.60	1.80	0	0
Washington	7	1.30	2.31	1	0	–	–	–	–	–
York	17	1.30	1.51	0	0	4	4.71	7.72	2	0
Maryland	2595	2.02	3.62	556	46	1018	4.20	9.34	406	83
Anne Arundel	103	1.40	2.10	10	0	23	2.02	2.70	5	0
Baltimore	113	1.80	3.14	23	2	62	2.63	4.91	19	3
Calvert	28	1.70	2.24	6	0	4	5.74	10.42	1	1

Carroll*	79	4.50	8.65	35	6	52	11.50	22.40	30	13
Cecil	5	0.95	1.10	0	0	–	–	–	–	–
Charles	26	1.20	1.50	1	0	6	3.25	3.43	2	0
Frederick	170	3.40	6.20	68	9	129	6.80	13.30	65	22
Garrett	5	2.00	2.90	2	0	3	17.90	19.72	1	2
Harford	15	2.50	5.22	3	1	10	3.20	4.22	5	0
Howard	281	3.50	5.42	121	10	108	7.20	10.00	72	10
Montgomery	1398	1.90	2.90	240	9	522	3.60	7.32	188	26
Prince Geo.	233	1.32	2.34	21	5	3	0.50	0.53	0	0
Prince Geo.†	47	1.03	1.31	1	0	64	1.94	3.70	10	2
Queen Anne	4	0.65	0.70	0	0	–	–	–	–	–
St. Mary's	7	1.23	1.50	0	0	–	–	–	–	–
Talbot	10	1.20	1.80	1	0	–	–	–	–	–
Washington	27	5.20	20.74	13	2	8	14.12	97.95	4	2
Massachusetts	658	1.10	1.73	46	4	219	2.52	4.64	48	5
Barnstable	13	0.82	1.00	0	0	–	–	–	–	–
Berkshire	45	0.80	1.00	0	0	6	1.54	1.80	0	0
Bristol	19	1.51	4.90	2	1	4	1.35	1.65	0	0
Essex	72	1.30	1.85	10	0	15	3.50	5.82	5	1
Franklin	12	1.23	1.40	0	0	–	–	–	–	–
Hampden	63	0.80	1.03	1	0	17	1.64	2.10	2	0
Middlesex	220	1.20	1.80	18	1	82	2.93	4.42	22	2
Norfolk	79	0.90	1.13	2	0	32	1.82	2.70	3	0
Plymouth	24	1.01	1.54	1	0	8	1.64	2.01	0	0
Suffolk	18	0.90	1.13	1	0	10	2.90	4.75	4	0
Worcester	56	1.45	2.72	10	1	17	4.01	5.41	6	1
Michigan	512	1.30	1.80	38	0	148	1.80	2.45	17	1
Allegan	12	0.81	1.00	0	0	6	1.40	1.65	0	0
Barry	8	1.33	2.04	1	0	–	–	–	–	–
Berrien	7	0.90	1.10	0	0	3	1.90	2.20	0	0
Berrien†	45	1.00	1.20	0	0	–	–	–	–	–
Calhoun	10	1.50	2.10	2	0	4	3.85	4.40	2	0

State/County	Living Areas					Basements				
	NMBR	Mean	Aver.	4–20	>20	NMBR	Mean	Aver.	4–20	>20
Genessee	13	0.82	1.04	0	0	4	2.00	2.50	1	0
Houghton	4	1.00	2.70	1	0	–	–	–	–	–
Ingham	19	1.34	1.65	1	0	4	1.15	1.44	0	0
Ionia	4	1.30	1.80	0	0	3	2.61	2.70	0	0
Jackson	5	1.30	1.60	0	0	–	–	–	–	–
Kalamazoo	26	1.73	2.43	5	0	7	1.70	2.33	1	0
Kalamazoo†	26	1.33	1.90	2	0	–	–	–	–	–
Kent	105	1.20	1.64	4	0	32	1.53	2.44	3	1
Lapeer	7	2.11	3.60	3	0	–	–	–	–	–
Livingston	4	1.40	1.95	0	0	–	–	–	–	–
Macomb*	13	1.10	1.32	0	0	–	–	–	–	–
Marquette	9	1.34	1.60	1	0	6	2.10	3.40	1	0
Mecosta	4	0.62	0.71	0	0	–	–	–	–	–
Monroe*	11	1.30	1.60	1	0	4	1.80	2.00	0	0
Oakland	71	1.30	1.62	3	0	16	1.64	1.90	1	0
Ottawa	20	1.34	1.90	1	0	8	1.31	1.70	0	0
St. Clair	3	0.93	1.30	0	0	–	–	–	–	–
St. Clair†	48	0.72	0.91	1	0	–	–	–	–	–
Van Buren	4	0.75	0.92	0	0	3	1.60	2.61	1	0
Washtenau	38	2.10	2.70	6	0	19	2.12	2.52	2	0
Washtenau†	61	1.71	3.10	10	1	–	–	–	–	–
Wayne	54	1.13	1.51	2	0	11	1.50	1.84	1	0
Minnesota	170	2.10	3.12	37	1	41	3.60	5.20	19	1
Blue Earth	6	4.00	5.00	3	0	–	–	–	–	–
Carver*	6	1.93	2.20	0	0	–	–	–	–	–
Clay	7	2.00	3.15	2	0	–	–	–	–	–
Hennepin	40	2.00	2.72	8	0	9	2.93	3.94	4	0
Morrison	8	1.04	1.23	0	0	–	–	–	–	–
Pine	4	1.30	1.35	0	0	–	–	–	–	–
Ramsey	15	1.72	3.22	3	0	4	3.10	4.00	1	0

Stearns*	7	2.80	3.61	4	0	—	—	—	—	—
St. Louis	20	1.94	2.70	3	0	9	2.50	3.05	4	0
Washington	7	2.50	3.02	1	0	—	—	—	—	—
Mississippi	14	2.12	2.40	1	0	—	—	—	—	—
Harrison†	15	1.00	1.10	0	0	—	—	—	—	—
Missouri	241	1.42	2.03	26	1	66	2.42	3.40	14	1
Boone	4	1.04	1.33	0	0	—	—	—	—	—
Cass	4	1.40	1.70	0	0	—	—	—	—	—
Clay*	5	1.70	2.00	0	0	—	—	—	—	—
Jackson	35	2.30	3.44	9	1	5	2.75	3.50	1	0
Jasper	18	0.90	1.10	0	0	—	—	—	—	—
Jefferson	6	1.30	1.50	0	0	—	—	—	—	—
Jefferson†	25	1.04	1.44	1	0	—	—	—	—	—
Platte	6	1.40	2.10	2	0	—	—	—	—	—
St. Charles*	9	1.80	2.04	1	0	8	2.30	2.80	1	0
St. Francois	8	1.35	1.81	1	0	5	2.14	2.63	1	0
St. Louis	87	1.40	1.90	8	0	30	2.40	3.10	9	0
Montana	41	2.80	3.90	14	0	8	2.50	3.00	3	0
Flathead	4	1.93	2.12	0	0	—	—	—	—	—
Ravalli	4	1.40	1.80	0	0	—	—	—	—	—
Yellowstone	4	3.83	5.80	2	0	—	—	—	—	—
Nevada	24	2.20	3.00	3	0	—	—	—	—	—
Clark	4	1.35	1.40	0	0	—	—	—	—	—
Clark†	33	1.10	1.41	1	0	—	—	—	—	—
Douglas	4	1.85	1.91	0	0	—	—	—	—	—
Washoe	12	2.75	4.10	3	0	—	—	—	—	—
New Hampshire	301	1.70	2.92	46	4	53	2.75	3.93	17	0
Belknap	9	1.00	1.10	0	0	—	—	—	—	—
Carroll	12	4.10	5.11	7	0	—	—	—	—	—
Cheshire	40	1.52	2.44	5	0	7	3.10	4.20	2	0
Coos	10	1.31	3.50	0	1	—	—	—	—	—
Grafton	25	1.54	2.70	4	0	3	3.24	3.73	1	0

State/County	Living Areas					Basements				
	NMBR	Mean	Aver.	4–20	>20	NMBR	Mean	Aver.	4–20	>20
Hillsboro	87	1.71	3.21	12	2	22	3.41	4.50	10	0
Merrimac	21	1.23	2.02	3	0	3	2.01	2.10	0	0
Rockingham	57	1.62	2.55	7	0	6	1.50	1.70	0	0
Strafford	19	3.00	4.81	5	1	–	–	–	–	–
Sullivan	8	1.30	1.55	1	0	5	4.52	6.35	3	0
New Jersey	12217	1.45	3.35	1580	234	3776	3.10	7.00	1173	227
Atlantic†	33	0.60	0.65	0	0	–	–	–	–	–
Bergen	832	0.75	1.01	22	1	244	1.50	1.90	20	0
Burlington	23	0.90	1.01	0	0	–	–	–	–	–
Camden	21	1.43	1.91	3	0	7	2.70	3.10	1	0
Essex	317	0.85	1.42	11	2	134	1.42	2.05	5	1
Gloucester	7	1.11	1.30	0	0	–	–	–	–	–
Hudson	40	0.81	1.30	2	0	6	1.55	2.10	1	0
Hunterdon	1384	2.83	10.00	374	94	481	6.42	16.22	218	79
Mercer	215	1.53	3.10	30	4	103	4.05	8.70	37	9
Middlesex	189	1.00	1.30	9	0	51	2.40	5.54	15	2
Monmouth	76	1.04	1.61	5	0	11	2.00	3.71	3	0
Monmouth†	73	0.70	0.94	1	0	–	–	–	–	–
Morris	4493	1.35	2.54	497	60	1488	2.94	6.00	491	59
Morris†	142	1.20	1.80	5	1	–	–	–	–	–
Ocean	19	0.72	0.93	0	0	–	–	–	–	–
Passaic	833	1.03	1.52	54	0	238	2.10	3.30	39	6
Somerset	678	1.55	3.40	95	20	269	3.60	7.64	78	26
Sussex	1935	1.70	2.75	288	16	405	3.40	6.74	155	21
Union	207	0.90	1.73	5	3	70	1.71	3.70	8	2
Warren	481	2.61	5.40	123	23	155	5.41	9.90	76	21
New Mexico	49	2.64	4.80	11	3	–	–	–	–	–
Bernalillo	13	3.70	5.42	5	1	–	–	–	–	–
Santa Fe*	12	2.81	3.42	4	0	–	–	–	–	–
Socorro	4	2.62	2.94	1	0	–	–	–	–	–

New York	2202	1.20	2.03	223	14	614	2.50	4.54	151	26
Albany	28	0.90	1.32	2	0	8	4.50	5.70	4	0
Allegany	4	1.40	3.11	1	0	–	–	–	–	–
Bronx	71	0.85	1.30	5	0	19	1.40	1.90	2	0
Cattaraugus	6	4.70	8.00	3	1	3	5.60	10.45	0	1
Cayuga	5	1.50	1.90	0	0	4	1.61	2.20	0	0
Chatauqua	12	1.65	5.10	0	2	4	4.40	5.00	3	0
Chatauqua†	62	1.20	2.00	10	0	–	–	–	–	–
Chemung*	20	1.73	2.52	4	0	5	9.70	12.85	2	2
Columbia	17	2.01	3.03	4	0	5	6.55	13.31	3	1
Cortland	7	5.80	10.23	4	1	–	–	–	–	–
Delaware	4	1.95	2.82	1	0	–	–	–	–	–
Dutchess	111	1.65	2.61	22	1	30	2.85	3.52	10	0
Erie	63	1.00	2.20	6	1	16	1.61	2.65	4	0
Essex	5	0.75	1.50	1	0	–	–	–	–	–
Franklin	4	0.41	0.40	0	0	–	–	–	–	–
Fulton	6	1.11	1.81	1	0	–	–	–	–	–
Genesee	8	1.41	1.55	0	0	–	–	–	–	–
Greene	11	1.60	2.45	3	0	–	–	–	–	–
Hamilton	4	0.50	0.51	0	0	–	–	–	–	–
Jefferson	11	1.00	1.14	0	0	–	–	–	–	–
Kings	22	0.62	0.70	0	0	6	1.93	2.15	0	0
Lewis	4	1.94	2.80	1	0	–	–	–	–	–
Livingston	7	1.70	4.10	2	0	–	–	–	–	–
Madison	7	1.50	2.14	2	0	3	8.85	10.20	3	0
Manhattan	6	0.41	0.42	0	0	–	–	–	–	–
Monroe*	29	1.10	1.25	1	0	10	1.44	1.72	0	0
Nassau	56	0.64	0.80	0	0	13	0.84	1.00	0	0
New York	9	0.85	1.40	1	0	5	1.30	1.72	1	0
Niagara	4	1.10	1.40	0	0	–	–	–	–	–
Oneida	18	1.54	2.70	2	0	6	5.22	8.95	2	1
Oneida†	59	1.22	2.50	5	1	–	–	–	–	–

State/County	Living Areas					Basements				
	NMBR	Mean	Aver.	4–20	>20	NMBR	Mean	Aver.	4–20	>20
Onondaga	94	2.24	4.83	26	1	57	4.61	9.03	22	7
Ontario	5	2.11	2.20	0	0	–	–	–	–	–
Orange	179	1.33	2.00	17	0	48	2.40	3.74	10	1
Oswego	5	0.73	0.84	0	0	4	0.65	0.70	0	0
Otsego	6	1.91	2.50	1	0	–	–	–	–	–
Putnam	372	1.80	3.10	66	6	77	2.82	4.30	23	2
Queens	39	0.54	0.61	0	0	4	1.70	2.12	1	0
Rensselaer	22	1.50	1.75	1	0	6	5.10	6.50	3	0
Richmond	33	0.74	0.85	0	0	5	0.82	1.22	0	0
Rockland	149	0.90	1.20	3	0	41	1.64	2.13	4	0
Saratoga	14	1.45	2.10	3	0	4	5.73	7.92	3	0
Schenectady	16	0.70	0.73	0	0	6	1.73	3.20	1	0
Steuben	5	1.10	1.25	0	0	–	–	–	–	–
St. Lawrence	12	0.80	1.10	1	0	3	3.25	5.64	2	0
Suffolk	51	0.70	0.73	0	0	9	1.50	1.90	1	0
Sullivan	15	0.74	1.01	1	0	5	3.80	5.60	2	0
Tioga	16	2.40	3.14	5	0	5	6.82	8.25	4	0
Tompkins	11	1.32	1.93	2	0	6	2.80	5.90	2	1
Ulster	56	1.23	1.62	2	0	15	2.80	4.10	4	0
Warren	10	0.70	0.80	0	0	–	–	–	–	–
Washington	5	1.62	2.81	1	0	–	–	–	–	–
Westchester	417	1.02	1.43	20	1	130	2.23	4.40	15	8
North Carolina	292	2.00	3.21	54	4	29	2.70	3.80	9	0
Buncombe	16	2.60	3.60	4	0	–	–	–	–	–
Durham	8	1.55	1.90	0	0	–	–	–	–	–
Forsythe	12	4.21	7.13	5	1	–	–	–	–	–
Franklin	7	2.70	3.32	2	0	–	–	–	–	–
Gaston	13	3.50	4.72	5	0	–	–	–	–	–
Guilford	16	1.20	1.55	2	0	–	–	–	–	–
Henderson	44	4.10	6.05	22	1	4	8.00	9.00	4	0

Jackson	7	1.11	1.24	0	0	–	–	–	–	–
Macon	18	2.01	2.35	2	0	6	2.12	2.40	1	0
Mecklenberg	9	0.75	0.90	0	0	–	–	–	–	–
Moore	6	1.62	1.80	0	0	–	–	–	–	–
Orange	5	1.40	1.65	0	0	–	–	–	–	–
Union	4	0.90	1.10	0	0	–	–	–	–	–
Wake	38	1.51	1.95	3	0	4	2.10	3.73	1	0
Watauga	5	3.62	5.12	2	0	–	–	–	–	–
North Dakota	47	2.34	3.61	15	0	9	3.81	5.63	4	0
North Dakota†	159	2.72	5.20	5.4	3	–	–	–	–	–
Burleigh†	17	1.85	2.30	1	0	–	–	–	–	–
Cass	17	2.30	4.10	6	0	4	3.30	4.61	2	0
Cass†	10	1.80	2.50	1	0	–	–	–	–	–
Grand Forks	3	7.50	8.53	3	0	–	–	–	–	–
Grand Forks†	77	3.60	7.31	36	3	–	–	–	–	–
Morton†	5	1.00	1.30	0	0	–	–	–	–	–
Stutsman	4	2.10	2.71	1	0	–	–	–	–	–
Stutsman†	4	2.30	2.85	2	0	–	–	–	–	–
Walsh†	4	4.85	6.80	3	0	–	–	–	–	–
Ohio	518	2.20	3.64	128	7	183	3.45	6.70	64	11
Allen	4	1.81	3.50	1	0	–	–	–	–	–
Ashtabula*	4	0.90	0.91	0	0	–	–	–	–	–
Athens	4	1.60	1.80	0	0	–	–	–	–	–
Belmont	4	3.35	4.73	2	0	–	–	–	–	–
Butler	6	4.10	4.92	3	0	–	–	–	–	–
Clark	5	1.30	1.81	1	0	3	12.20	13.21	2	1
Clermont	4	1.10	1.23	0	0	3	8.20	8.34	3	0
Crawford	5	1.20	1.90	1	0	–	–	–	–	–
Cuyahoga	65	1.24	2.60	2	1	41	1.45	1.90	2	0
Delaware	11	2.53	3.50	4	0	–	–	–	–	–
Erie	5	0.90	0.95	0	0	–	–	–	–	–
Fairfield	5	2.55	3.50	2	0	–	–	–	–	–

State/County	Living Areas					Basements				
	NMBR	Mean	Aver.	4–20	>20	NMBR	Mean	Aver.	4–20	>20
Franklin	57	4.70	6.60	29	3	19	10.43	16.02	12	4
Geauga	11	0.80	1.11	1	0	4	0.90	0.94	0	0
Greene	11	4.72	6.30	6	0	10	3.93	5.30	5	0
Hamilton	22	1.84	3.04	4	0	5	2.12	6.50	0	1
Henry	4	3.80	4.85	2	0	–	–	–	–	–
Holmes	4	11.60	12.20	4	0	–	–	–	–	–
Lake	10	1.12	1.70	1	0	–	–	–	–	–
Licking	9	3.91	5.04	4	0	–	–	–	–	–
Lorain	6	1.95	2.50	1	0	6	4.80	5.71	5	0
Lucas	13	1.40	1.60	0	0	6	6.54	15.10	2	1
Mahoning	6	2.10	4.64	0	1	–	–	–	–	–
Medina	15	2.44	3.00	4	0	4	2.90	5.14	1	0
Miami	22	2.63	4.00	8	0	–	–	–	–	–
Montgomery*	67	2.65	4.10	17	2	30	4.22	6.85	12	2
Portage	5	0.72	0.81	0	0	–	–	–	–	–
Richland	5	3.30	4.00	1	0	–	–	–	–	–
Stark	14	2.01	3.02	3	0	3	1.90	1.92	0	0
Summit	32	1.80	2.32	5	0	6	1.80	2.25	1	0
Trumbull	10	1.82	3.40	2	0	4	1.60	1.70	0	0
Washington	4	1.21	1.34	0	0	–	–	–	–	–
Wood	5	1.24	1.40	0	0	–	–	–	–	–
Oklahoma	52	1.32	1.71	3	0	–	–	–	–	–
Oklahoma	9	1.55	1.80	0	0	–	–	–	–	–
Tulsa	9	1.31	1.40	0	0	–	–	–	–	–
Tulsa†	22	1.01	1.30	0	0	–	–	–	–	–
Oregon	83	1.42	2.60	10	1	17	1.31	1.95	2	0
Clackamas*	5	1.30	1.61	0	0	–	–	–	–	–
Jackson	14	2.41	5.24	3	1	3	2.05	3.71	1	0
Lane	14	0.62	0.80	0	0	–	–	–	–	–
Lincoln	5	1.12	1.41	0	0	–	–	–	–	–

Marion†	45	1.30	1.60	1	0	–	–	–	–	–
Multnomah	12	1.82	2.11	1	0	5	1.14	1.33	0	0
Washington*	7	1.23	1.32	0	0	–	–	–	–	–
Pennsylvania	4744	2.94	7.00	1450	327	1962	6.70	15.80	942	336
Adams	8	2.20	4.92	1	1	4	1.60	1.82	0	0
Allegheny	370	1.80	3.14	74	4	54	3.15	4.34	18	1
Armstrong	15	3.10	4.05	7	0	4	3.70	5.15	1	0
Beaver	26	5.01	13.14	10	3	3	9.94	25.41	1	1
Bedford	5	1.60	2.30	2	0	5	5.85	9.03	3	0
Berks	281	4.30	9.14	105	37	231	5.10	11.00	92	34
Berks†	46	2.02	2.91	11	0	–	–	–	–	–
Blair	11	1.83	2.62	3	0	–	–	–	–	–
Bradford	13	1.62	2.40	1	0	–	–	–	–	–
Bucks	199	2.30	3.80	40	7	78	4.65	10.14	34	10
Butler	24	3.50	7.40	5	3	4	3.80	10.41	0	1
Cambria	15	1.90	4.65	1	2	–	–	–	–	–
Cambria†	84	1.30	3.00	7	2	–	–	–	–	–
Carbon	18	2.40	7.03	3	3	6	6.42	14.10	3	1
Centre	29	7.05	9.40	17	4	5	11.70	13.43	4	1
Chester	199	2.13	5.00	34	7	70	5.53	18.30	26	9
Clarion	5	2.70	3.44	1	0	3	0.35	0.32	0	0
Clearfield	11	1.84	2.80	2	0	–	–	–	–	–
Columbia	11	2.40	3.80	2	0	–	–	–	–	–
Crawford	6	0.90	1.10	0	0	6	2.30	3.30	2	0
Cumberland	175	5.50	8.94	97	15	111	12.30	20.84	57	38
Cumberland†	38	4.30	6.80	21	2	–	–	–	–	–
Dauphin	75	4.53	8.10	36	7	33	7.60	11.90	23	2
Delaware	85	1.23	1.82	7	0	25	2.02	3.03	3	1
Erie	19	1.70	3.93	4	1	19	2.20	5.72	5	1
Fayette	7	2.40	4.10	2	0	19	2.20	5.72	5	1
Franklin	61	3.23	5.52	22	3	13	11.23	19.05	8	3
Franklin†	56	2.40	3.71	16	0	–	–	–	–	–

State/County	Living Areas					Basements				
	NMBR	Mean	Aver.	4–20	>20	NMBR	Mean	Aver.	4–20	>20
Greene	4	1.60	1.72	0	0	–	–	–	–	–
Huntington	4	1.10	1.30	0	0	–	–	–	–	–
Indiana	13	1.00	1.60	2	0	–	–	–	–	–
Jefferson	6	2.11	2.40	1	0	–	–	–	–	–
Lackawanna	9	1.40	1.60	0	0	–	–	–	–	–
Lancaster	443	4.55	9.00	187	51	221	10.00	18.10	119	60
Lawrence	7	2.00	2.20	0	0	–	–	–	–	–
Lebanon	121	4.05	7.50	46	13	65	11.45	21.50	37	15
Lebanon†	64	3.10	6.80	21	4	–	–	–	–	–
Lehigh	872	3.50	7.50	338	63	424	8.03	15.90	234	77
Lehigh†	76	2.90	5.80	22	5	–	–	–	–	–
Luzerne	35	1.50	2.14	4	0	6	8.23	15.62	4	1
Lycoming	14	4.64	27.64	2	4	5	11.11	17.70	2	2
Lycoming†	80	1.73	2.93	13	1	–	–	–	–	–
Mercer	11	1.10	2.43	1	0	–	–	–	–	–
Montgomery*	352	1.90	4.32	40	16	149	3.30	5.72	49	8
Northampton*	651	3.71	11.80	236	55	275	9.30	29.30	157	55
Northampton†	91	2.00	4.00	19	2	–	–	–	–	–
Northmbrlnd.	3	0.60	0.60	0	0	–	–	–	–	–
Northmbrlnd.†	46	1.40	2.10	4	0	–	–	–	–	–
Perry	7	6.10	8.60	4	1	5	10.80	40.70	1	2
Philadelphia	35	1.10	1.54	4	0	12	2.93	5.70	4	0
Pike	46	1.31	1.95	7	0	16	2.22	4.00	4	1
Schuylkill	16	4.31	10.23	5	1	3	5.10	16.13	0	1
Schuylkill†	150	2.00	3.44	38	3	–	–	–	–	–
Somerset	6	4.83	18.53	1	2	–	–	–	–	–
Sullivan	6	1.60	1.70	0	0	–	–	–	–	–
Susquehanna	7	1.42	1.73	0	0	–	–	–	–	–
Tioga	4	1.02	1.11	0	0	–	–	–	–	–
Union	9	1.54	3.55	2	0	–	–	–	–	–

Venango	5	1.22	2.90	1	0	–	–	–	–	–
Warren	15	1.00	1.10	0	0	–	–	–	–	–
Washington	45	1.70	2.94	4	2	–	–	–	–	–
Wayne	37	1.34	2.44	3	1	5	2.20	2.33	0	0
Westmoreland	61	1.90	3.20	17	0	9	3.04	3.40	3	0
Wyoming	6	2.23	2.54	1	0	–	–	–	–	–
York	99	3.70	8.74	28	14	34	7.70	11.61	21	4
Rhode Island	73	1.25	1.90	8	0	16	2.71	3.32	7	0
Bristol	4	1.01	1.71	1	0	–	–	–	–	–
Kent	6	0.80	1.22	1	0	3	1.14	1.20	0	0
Newport	23	0.94	1.20	1	0	3	4.30	4.50	2	0
Newport†	35	0.75	1.03	1	0	–	–	–	–	–
Providence	15	1.23	1.60	1	0	4	1.90	2.40	1	0
Washington	13	2.63	3.80	3	0	–	–	–	–	–
So. Carolina	55	1.20	1.71	5	0	4	6.00	9.54	3	0
Charleston	8	0.70	0.80	0	0	–	–	–	–	–
Charleston†	31	0.80	1.11	1	0	–	–	–	–	–
Georgetown	4	1.01	1.11	0	0	–	–	–	–	–
Greenville*	7	2.61	4.04	3	0	–	–	–	–	–
Horry	4	1.15	1.24	0	0	–	–	–	–	–
Spartanburg†	19	1.31	1.70	2	0	–	–	–	–	–
South Dakota	17	1.95	2.60	5	0	4	2.82	2.82	0	0
South Dakota†	86	2.60	3.91	26	0	–	–	–	–	–
Brookings	5	3.02	3.80	3	0	–	–	–	–	–
Codington†	5	2.24	2.64	0	0	–	–	–	–	–
Hughes†	6	2.04	2.32	1	0	–	–	–	–	–
Kingsbury†	4	4.40	4.95	1	0	–	–	–	–	–
Minnehaha†	14	2.61	3.05	2	0	–	–	–	–	–
Pennington†	7	4.40	6.40	3	0	–	–	–	–	–
Tennessee	854	2.35	3.60	208	11	120	4.10	8.10	49	12
Anderson	15	3.00	4.40	5	0	–	–	–	–	–
Blount	9	3.60	3.95	3	0	–	–	–	–	–

State/County	Living Areas					Basements				
	NMBR	Mean	Aver.	4–20	>20	NMBR	Mean	Aver.	4–20	>20
Bradley	7	1.40	1.60	0	0	–	–	–	–	–
Coffee	12	1.53	2.30	2	0	–	–	–	–	–
Cumberland	4	0.80	1.00	0	0	–	–	–	–	–
Davidson	233	2.40	3.50	65	0	48	5.10	11.30	21	7
Dickson	4	1.73	1.81	0	0	–	–	–	–	–
Franklin	7	1.50	1.70	0	0	–	–	–	–	–
Hamilton*	44	1.65	2.05	5	0	11	2.15	2.60	2	0
Jefferson	4	4.00	4.21	2	0	–	–	–	–	–
Knox	42	2.10	3.02	10	0	4	4.13	4.43	3	0
Maury	5	2.40	2.60	1	0	–	–	–	–	–
Montgomery*	6	2.25	2.95	1	0	–	–	–	–	–
Putnam	7	2.10	3.60	2	0	–	–	–	–	–
Roane	10	2.80	6.70	4	1	–	–	–	–	–
Robertson	7	1.50	1.70	0	0	–	–	–	–	–
Rutherford	18	1.80	2.23	2	0	3	3.50	4.70	2	0
Shelby	5	1.20	1.30	0	0	–	–	–	–	–
Sullivan	10	5.20	6.41	7	0	3	9.94	11.60	2	1
Sumner	21	2.11	2.94	3	0	3	6.10	13.03	0	1
Washington	14	3.25	4.40	5	0	–	–	–	–	–
White	6	4.01	14.63	2	1	–	–	–	–	–
Williamson*	311	2.51	3.60	80	5	25	3.15	6.20	10	2
Texas	121	1.80	2.83	18	1	6	3.90	5.83	2	0
Bexar	6	0.70	0.80	0	0	–	–	–	–	–
Dallas	18	3.00	4.31	6	0	–	–	–	–	–
Galveston†	13	0.45	0.43	0	0	–	–	–	–	–
Harris	13	1.10	1.40	0	0	–	–	–	–	–
Jefferson†	23	0.55	0.60	0	0	–	–	–	–	–
Nueces	3	0.44	0.40	0	0	–	–	–	–	–
Nueces†	18	0.43	0.40	0	0	–	–	–	–	–
Tarrant	10	2.90	4.22	5	0	–	–	–	–	–

Travis	9	1.30	1.52	0	0	–	–	–	–	–
Williamson*	5	1.84	2.11	1	0	–	–	–	–	–
Utah	30	1.84	2.71	5	0	4	1.64	4.00	1	0
Utah†	125	1.42	1.93	13	0	–	–	–	–	–
Davis†	18	1.40	2.20	5	0	–	–	–	–	–
Salt Lake	20	2.30	3.33	5	0	–	–	–	–	–
Salt Lake†	87	1.40	1.80	5	0	–	–	–	–	–
Utah†	5	2.50	3.04	1	0	–	–	–	–	–
Weber	5	1.30	1.43	0	0	–	–	–	–	–
Weber†	4	0.70	0.80	0	0	–	–	–	–	–
Vermont	72	1.00	1.30	2	0	17	2.40	3.90	6	0
Bennington	4	1.73	2.00	0	0	3	2.54	2.90	1	0
Chittenden	8	0.75	0.93	0	0	–	–	–	–	–
Franklin	5	0.70	0.71	0	0	–	–	–	–	–
Rutland	13	0.83	1.15	0	0	–	–	–	–	–
Windsor	22	0.95	1.25	0	0	4	5.20	8.20	3	0
Virginia	1528	1.60	2.50	203	10	510	2.83	5.10	149	13
Albemarle	9	1.80	5.20	1	1	–	–	–	–	–
Arlington	82	1.22	1.70	5	0	40	1.80	2.34	5	0
Augusta	5	2.92	3.31	2	0	–	–	–	–	–
Bedford	7	1.03	1.15	0	0	–	–	–	–	–
Campbell	7	3.30	5.10	3	0	–	–	–	–	–
Chesterfield	17	1.82	3.00	1	1	–	–	–	–	–
Clarke	5	3.90	4.80	3	0	–	–	–	–	–
Fairfax	959	1.71	2.65	140	7	331	2.90	5.60	95	11
Fairfax†	89	1.40	2.00	11	0	–	–	–	–	–
Fauquer	27	1.90	3.74	7	0	11	2.33	3.41	3	0
Frederick	10	3.20	5.70	3	1	5	6.53	7.72	4	0
Henrico*	10	1.64	2.10	2	0	–	–	–	–	–
Loudon*	52	1.50	2.22	10	0	21	4.10	5.74	11	0
Montgomery	6	3.05	4.23	2	0	–	–	–	–	–
Pittsylvania	13	2.40	2.92	2	0	–	–	–	–	–

State/County	Living Areas					Basements				
	NMBR	Mean	Aver.	4–20	>20	NMBR	Mean	Aver.	4–20	>20
Prince ———	84	1.40	1.80	6	0	36	3.00	4.54	11	1
Rappahannock	5	1.12	1.30	0	0	–	–	–	–	–
Richmond	9	1.90	4.00	2	0	–	–	–	–	–
Roanoke	17	2.30	3.00	5	0	8	2.14	3.20	2	0
Rockingham*	12	1.75	2.05	1	0	–	–	–	–	–
Spotsylvania	5	1.10	1.15	0	0	–	–	–	–	–
Stafford	9	1.43	1.74	1	0	–	–	–	–	–
Warren	4	1.55	1.60	0	0	–	–	–	–	–
York	7	0.72	0.80	0	0	–	–	–	–	–
Washington	221	1.80	5.34	36	16	35	3.42	7.50	9	4
Benton	7	1.50	1.60	0	0	–	–	–	–	–
Clark*	7	1.30	1.64	0	0	–	–	–	–	–
Douglas	8	1.20	1.35	0	0	–	–	–	–	–
King	18	0.70	1.03	1	0	–	–	–	–	–
Okanogan	8	7.40	26.11	1	3	–	–	–	–	–
Pierce	22	0.91	2.22	2	0	–	–	–	–	–
Snohomish	8	0.70	0.80	0	0	–	–	–	–	–
Spokane	85	4.21	8.40	29	11	23	5.73	10.30	7	4
Stevens	6	3.13	5.41	1	1	–	–	–	–	–
Whatcom	19	0.40	0.40	0	0	–	–	–	–	–
Yakima	5	1.30	1.65	0	0	–	–	–	–	–
West Virginia	99	2.00	3.94	16	6	16	7.60	11.00	11	2
Hampshire	4	1.33	2.80	1	0	–	–	–	–	–
Harrison	12	1.80	2.40	3	0	–	–	–	–	–
Jefferson	32	2.10	5.23	5	3	8	11.20	12.63	7	1
Marion*	6	1.90	2.60	1	0	–	–	–	–	–
Monongalia*	7	1.40	1.52	0	0	–	–	–	–	–
Morgan	6	3.23	9.33	0	2	–	–	–	–	–
Ohio	5	2.10	7.44	0	1	–	–	–	–	–
Wisconsin	171	1.80	3.14	26	6	45	3.34	4.92	15	2

Brown	9	1.30	1.72	1	0	3	2.63	3.35	1	0
Dane	7	1.50	1.82	1	0	—	—	—	—	—
Marathon*	12	1.80	4.34	2	1	4	6.20	9.32	1	1
Milwaukee	19	1.32	2.50	2	1	4	2.85	3.30	1	0
Oneida	4	1.02	1.20	0	0	—	—	—	—	—
Outagamie	7	1.52	1.80	0	0	—	—	—	—	—
Ozaukee	9	1.95	2.30	0	0	—	—	—	—	—
Portage	4	1.74	1.80	0	0	—	—	—	—	—
Racine	5	2.44	3.60	2	0	5	4.10	5.00	2	0
Sheboygan	4	2.15	2.90	2	0	—	—	—	—	—
Walworth	7	1.24	1.51	0	0	—	—	—	—	—
Washington*	4	1.20	2.02	1	0	—	—	—	—	—
Waukesha*	23	2.40	3.70	8	0	—	—	—	—	—
Wood	6	19.95	21.00	2	4	—	—	—	—	—
Wyoming	107	2.92	5.00	32	2	37	3.64	4.70	16	0
Wyoming†	77	1.90	2.63	11	1	—	—	—	—	—
Albany†	6	1.30	1.35	0	0	—	—	—	—	—
Big Horn†	4	1.80	2.02	0	0	—	—	—	—	—
Carbon	5	3.32	8.95	0	1	—	—	—	—	—
Carbon†	4	2.33	3.45	2	0	—	—	—	—	—
Converse	4	5.52	9.12	2	0	—	—	—	—	—
Converse†	6	2.75	4.22	2	0	—	—	—	—	—
Fremont	23	5.14	7.70	11	1	—	—	—	—	—
Fremont†	4	2.35	2.43	0	0	—	—	—	—	—
Laramie	7	1.85	2.20	0	0	7	2.22	2.40	0	0
Laramie†	13	1.60	2.12	3	0	—	—	—	—	—
Natrona	7	2.80	5.12	4	0	9	3.60	4.15	3	0
Natrona†	13	2.23	3.50	0	1	—	—	—	—	—
Park	5	0.80	1.02	0	0	—	—	—	—	—
Park†	4	2.24	2.40	0	0	—	—	—	—	—
Sheridan	23	2.50	3.50	8	0	8	7.30	7.71	8	0
Teton	17	2.95	5.15	5	0	—	—	—	—	—
U.S.A. total	34280	1.70	3.64	5813	775	10338	3.57	8.21	3593	780

TABLE 3–B. ZIP CODES BY 5 DIGITS FOR NON-BASEMENTS (ONLY GROUPS WITH COUNTS > = 4 ARE PRINTED.)

Zip	Total	Mean	Avg.	–>4	4–>20	20–>	Zip	Total	Mean	Avg.	–>4	4–>20	20–>
00000	275	1.02	2.82	240	27	8	01915	6	1.20	1.62	5	1	0
01002	5	0.62	0.69	5	0	0	01940	5	1.40	1.70	4	1	0
01028	4	0.38	0.32	4	0	0	01944	7	2.31	3.72	5	2	0
01040	4	2.34	2.66	4	0	0	01945	5	0.67	0.73	5	0	0
01056	4	1.21	1.54	4	0	0	01966	8	0.74	0.85	8	0	0
01085	5	0.69	0.93	5	0	0	02025	4	0.78	0.99	4	0	0
01106	8	0.61	0.66	8	0	0	02030	7	0.82	1.05	7	0	0
01201	8	0.75	0.80	8	0	0	02052	6	1.92	2.22	5	1	0
01257	4	1.29	1.35	4	0	0	02090	5	0.63	0.69	5	0	0
01262	12	0.64	0.74	12	0	0	02114	5	1.30	1.45	5	0	0
01267	4	0.71	0.89	4	0	0	02131	4	0.81	0.99	4	0	0
01450	6	1.61	2.22	5	1	0	02139	4	0.85	0.99	4	0	0
01451	14	3.25	5.84	8	5	1	02148	8	0.54	0.58	8	0	0
01701	7	0.70	0.80	7	0	0	02158	4	0.95	1.15	4	0	0
01720	9	1.04	1.68	7	2	0	02167	6	0.90	1.06	6	0	0
01741	11	3.25	4.73	9	1	1	02168	4	0.72	0.81	4	0	0
01742	8	2.50	3.76	6	2	0	02169	4	0.54	0.58	4	0	0
01748	6	1.35	1.61	6	0	0	02173	9	1.45	1.80	9	0	0
01752	4	1.60	1.70	4	0	0	02174	5	0.77	0.88	5	0	0
01757	6	0.90	1.10	6	0	0	02176	6	0.47	0.47	6	0	0
01760	6	0.97	1.34	6	0	0	02181	9	0.80	1.07	9	0	0
01770	4	1.13	1.20	4	0	0	02184	5	0.79	0.88	5	0	0
01775	7	2.54	3.14	5	2	0	02193	6	2.15	2.23	6	0	0
01776	4	0.93	1.07	4	0	0	02194	4	0.74	0.83	4	0	0

01778	4	0.82	0.88	4	0	0	02818	5	1.52	2.04	4	1	0
01801	5	2.55	2.97	3	2	0	02835	6	1.02	1.39	5	1	0
01803	5	0.59	0.63	5	0	0	02840	10	0.73	0.93	10	0	0
01810	6	2.56	3.03	4	2	0	02879	4	4.08	5.32	2	2	0
01824	15	1.00	1.28	15	0	0	03031	7	1.28	1.43	7	0	0
01845	7	0.53	0.58	7	0	0	03034	4	3.16	3.68	3	1	0
01867	9	1.09	1.35	9	0	0	03049	5	1.86	2.11	5	0	0
01879	6	2.78	5.31	5	0	1	03051	4	1.43	1.54	4	0	0
01880	5	0.75	0.90	5	0	0	03053	6	1.26	1.56	6	0	0
01886	4	1.99	3.46	3	1	0	03062	6	1.76	2.13	5	1	0
01890	4	2.62	4.09	3	1	0	03087	10	2.17	3.42	8	2	0
01907	6	1.57	1.95	6	0	0	03102	14	2.26	6.30	10	3	1
03246	4	1.43	1.43	4	0	0	06413	6	0.92	1.01	6	0	0
03257	5	2.40	3.53	4	1	0	06417	4	1.17	1.18	4	0	0
03264	4	4.94	5.43	2	2	0	06426	4	1.06	1.14	4	0	0
03281	4	1.34	1.61	4	0	0	06430	10	0.80	1.30	9	1	0
03431	7	1.79	3.82	5	2	0	06437	7	0.96	1.36	7	0	0
03449	6	0.57	0.61	6	0	0	06441	4	1.30	1.64	4	0	0
03455	9	2.76	5.37	5	4	0	06443	4	3.09	4.09	2	2	0
03457	4	0.88	0.92	4	0	0	06447	5	2.02	2.16	5	0	0
03458	7	3.48	7.05	4	2	1	06460	7	1.15	1.65	6	1	0
03598	4	1.19	1.20	4	0	0	06470	4	0.95	1.26	4	0	0
03766	4	1.66	2.34	3	1	0	06477	5	2.78	4.25	4	1	0
03820	4	4.62	8.60	2	1	1	06480	4	1.35	1.90	4	0	0
03824	9	2.57	4.03	7	2	0	06483	4	1.44	2.92	3	1	0
03833	4	0.67	0.86	4	0	0	06484	7	1.15	1.69	7	0	0
03857	4	1.92	2.13	4	0	0	06488	4	1.08	1.40	4	0	0
04011	11	1.75	3.47	9	2	0	06489	4	0.77	0.91	4	0	0
04015	11	2.12	2.51	9	2	0	06525	14	1.35	2.09	12	2	0
04055	4	2.31	2.76	3	1	0	06611	4	1.74	3.73	3	1	0
04071	8	4.46	7.56	6	1	1	06708	6	0.59	0.60	6	0	0
04103	4	0.71	0.76	4	0	0	06751	5	2.32	4.29	3	2	0

Zip	Total	Mean	Avg.	–>4	4–>20	20–>	Zip	Total	Mean	Avg.	–>4	4–>20	20–>
04210	12	1.20	1.46	12	0	0	06759	4	1.14	1.15	4	0	0
04240	8	0.92	1.54	7	1	0	06762	6	1.69	1.89	6	0	0
04342	4	0.66	1.83	3	1	0	06776	6	1.53	3.80	5	1	0
04348	4	1.32	1.65	4	0	0	06784	8	0.57	0.62	8	0	0
04401	8	0.68	0.78	8	0	0	06791	5	1.44	1.61	5	0	0
04530	4	0.60	0.67	4	0	0	06793	6	1.18	1.26	6	0	0
04605	5	1.24	1.70	5	0	0	06801	5	0.91	0.91	5	0	0
04609	4	1.21	2.55	3	1	0	06810	6	0.72	0.98	6	0	0
04843	4	1.79	2.06	4	0	0	06811	14	0.39	0.29	14	0	0
04915	6	1.32	1.43	6	0	0	06812	22	0.63	0.71	22	0	0
06019	6	1.26	1.82	5	1	0	06820	5	2.33	2.43	5	0	0
06040	5	1.11	1.35	5	0	0	06830	9	1.09	1.29	9	0	0
06066	8	1.57	2.47	7	1	0	06831	5	1.51	1.63	5	0	0
06067	4	0.61	0.66	4	0	0	06840	7	1.33	2.11	5	2	0
06070	5	1.14	1.48	5	0	0	06877	19	1.12	1.54	18	1	0
06074	4	0.80	0.96	4	0	0	06880	20	1.88	5.04	16	3	1
06084	5	0.86	1.24	5	0	0	06883	7	3.20	3.84	4	3	0
06095	4	0.79	0.93	4	0	0	06896	10	1.52	2.80	7	3	0
06107	6	0.66	0.74	6	0	0	06897	10	2.47	2.72	8	2	0
06111	5	0.67	0.80	5	0	0	06902	4	1.49	4.88	3	1	0
06117	8	0.70	0.86	8	0	0	06903	14	1.31	1.75	13	1	0
06268	4	1.84	1.92	4	0	0	07003	23	0.64	0.75	23	0	0
06340	4	1.19	1.39	4	0	0	07005	134	1.09	1.37	132	2	0
06405	8	1.84	4.03	5	3	0	07006	29	0.69	0.91	28	1	0
07009	15	0.72	1.04	15	0	0	07094	4	1.17	1.31	4	0	0
07011	10	0.42	0.44	10	0	0	07095	4	1.12	1.22	4	0	0
07012	5	0.73	0.81	5	0	0	07109	7	0.50	0.53	7	0	0
07013	26	0.67	0.75	26	0	0	07110	11	1.07	1.37	11	0	0
07016	6	0.49	0.53	6	0	0	07111	4	2.60	7.60	3	0	1
07022	5	0.73	0.83	5	0	0	07114	6	2.55	3.00	4	2	0

07024	4	1.69	8.59	3	0	1	07204	4	0.46	0.47	4	0	0
07028	8	0.48	0.52	8	0	0	07302	6	0.77	0.80	6	0	0
07032	20	0.67	0.74	20	0	0	07400	5	0.73	1.28	4	1	0
07034	32	0.65	0.98	31	1	0	07401	13	0.67	0.80	13	0	0
07035	27	0.90	1.16	26	1	0	07403	61	0.84	1.18	58	3	0
07039	43	0.89	1.18	41	2	0	07405	165	1.23	1.88	148	16	1
07040	17	0.82	1.20	16	1	0	07407	6	0.70	0.77	6	0	0
07041	5	1.01	1.20	5	0	0	07410	59	0.67	0.91	57	2	0
07042	27	0.89	1.06	27	0	0	07416	41	1.83	2.32	38	3	0
07043	21	0.72	0.83	21	0	0	07417	29	0.96	1.33	27	2	0
07044	15	0.90	1.34	14	1	0	07418	49	2.37	3.35	37	12	0
07045	105	1.20	1.65	97	8	0	07419	43	2.06	2.97	33	10	0
07046	87	0.98	1.29	83	4	0	07420	21	1.40	1.76	20	1	0
07047	6	1.72	5.77	4	1	1	07421	72	1.82	2.85	61	10	1
07052	51	0.80	1.18	49	2	0	07422	154	1.94	3.12	128	24	2
07054	153	0.93	1.38	144	9	0	07423	6	1.33	1.81	6	0	0
07055	8	0.78	0.91	8	0	0	07424	24	0.60	0.72	24	0	0
07057	5	0.57	0.66	5	0	0	07428	15	2.16	2.73	12	3	0
07058	33	0.71	0.81	33	0	0	07430	111	0.86	1.23	109	2	0
07060	73	1.21	1.83	68	8	0	07432	5	1.12	1.59	4	1	0
07061	8	0.80	1.14	8	0	0	07435	40	1.77	3.05	32	8	0
07062	4	0.51	0.58	4	0	0	07436	80	0.80	1.22	78	1	1
07065	4	4.17	4.26	3	1	0	07438	112	2.13	3.90	84	26	2
07066	8	0.86	1.27	7	1	0	07439	24	2.29	3.12	20	4	0
07067	6	0.96	1.30	6	0	0	07440	8	0.77	0.87	8	0	0
07068	4	0.44	0.46	4	0	0	07442	34	1.22	1.82	29	5	0
07070	9	0.52	0.59	9	0	0	07444	35	1.21	1.72	33	2	0
07071	4	0.54	0.55	4	0	0	07446	57	0.54	0.62	57	0	0
07076	15	1.06	1.23	15	0	0	07450	43	0.81	1.01	43	0	0
07078	17	1.08	1.36	16	1	0	07452	24	062	0.68	24	0	0
07079	16	0.88	1.12	16	0	0	07456	220	1.17	1.65	210	9	1
07080	12	0.76	0.82	12	0	0	07457	12	1.06	1.26	12	0	0

Zip	Total	Mean	Avg.	–>4	4–>20	20–>
07081	11	0.59	0.65	11	0	0
07082	49	1.50	5.25	42	4	3
07083	12	0.96	1.19	12	0	0
07087	6	0.72	1.04	6	0	0
07090	28	0.81	0.94	28	0	0
07092	8	0.86	1.09	8	0	0
07470	135	0.83	1.09	132	3	0
07480	174	1.04	1.41	164	10	0
07481	48	0.87	1.06	48	0	0
07506	6	1.32	1.87	5	1	0
07508	17	0.52	0.57	17	0	0
07512	7	0.73	0.79	7	0	0
07601	4	1.48	1.89	4	0	0
07603	9	1.30	3.11	8	1	0
07605	9	1.75	3.12	6	3	0
07607	9	0.65	0.71	9	0	0
07620	7	1.87	2.62	5	2	0
07621	10	0.54	0.58	10	0	0
07624	7	0.45	0.44	7	0	0
07628	9	0.55	0.63	9	0	0
07630	7	0.67	0.80	7	0	0
07631	8	0.67	0.87	8	0	0
07632	9	1.06	1.49	8	1	0
07640	6	0.64	0.70	6	0	0
07641	10	2.30	3.54	7	3	0
07642	15	0.59	0.74	15	0	0
07644	9	0.52	0.57	9	0	0
07645	18	1.19	1.50	17	1	0
07646	10	0.49	0.49	10	0	0
07647	4	0.54	0.62	4	0	0
07458	49	0.77	1.07	48	1	0
07460	64	1.19	1.78	61	2	1
07461	266	1.64	2.87	216	46	4
07462	125	2.11	3.07	101	23	1
07463	6	0.77	0.92	6	0	0
07465	33	0.95	1.33	32	1	0
07822	15	2.31	3.19	11	4	0
07823	24	8.82	15.11	7	10	7
07825	41	2.15	3.67	31	10	0
07826	37	1.21	1.83	33	4	0
07827	4	1.50	1.74	4	0	0
07828	83	1.84	3.10	69	13	1
07830	137	2.77	5.70	87	41	9
07832	7	1.84	2.24	6	1	0
07834	216	1.06	1.38	208	8	0
07836	124	1.49	2.21	110	14	0
07838	17	5.04	9.73	8	8	1
07839	7	1.53	1.87	7	0	0
07840	188	2.33	4.28	135	46	7
07843	146	1.69	2.52	120	25	1
07844	15	7.87	16.34	2	10	3
07845	8	2.61	3.11	6	2	0
07847	11	1.09	1.44	11	0	0
07848	28	2.01	2.67	21	7	0
07849	76	1.51	2.32	65	10	1
07850	62	1.47	2.08	55	7	0
07852	5	1.49	1.80	5	0	0
07853	305	1.73	4.15	248	47	10
07856	19	1.33	1.51	19	0	0
07857	10	0.76	0.83	10	0	0

07649	11	0.60	0.68	11	0	0
07652	19	0.90	1.30	17	2	0
07656	11	1.07	1.21	11	0	0
07660	5	0.41	0.35	5	0	0
07661	11	0.57	0.60	11	0	0
07662	10	0.74	0.97	10	0	0
07666	27	0.66	0.89	26	1	0
07670	9	1.04	1.36	9	0	0
07675	32	0.86	1.16	30	2	0
07701	5	0.44	0.53	5	0	0
07726	6	0.48	0.50	6	0	0
07728	6	1.94	2.17	6	0	0
07730	4	1.01	1.24	4	0	0
07746	5	1.23	1.43	5	0	0
07747	4	0.68	0.72	4	0	0
07748	9	1.10	1.82	8	1	0
07801	135	1.60	2.88	108	24	3
07820	13	1.16	1.95	11	2	0
07821	123	1.51	2.28	108	15	0
07927	20	0.69	0.90	20	0	0
07928	105	1.25	2.00	91	14	0
07930	195	2.36	4.65	143	45	7
07931	37	1.73	4.01	28	8	1
07932	55	0.74	1.08	54	1	0
07933	16	1.03	1.33	15	1	0
07934	25	3.09	8.84	16	7	23
07936	29	0.66	0.79	29	0	0
07940	74	0.94	1.25	73	1	0
07945	242	1.63	3.59	192	45	5
07946	23	1.36	1.92	20	3	0
07950	227	1.26	3.64	207	18	2
07960	446	1.38	2.65	376	63	7
07863	21	2.18	2.56	18	3	0
07864	4	0.93	1.12	4	0	0
07865	15	0.34	2.98	11	4	0
07866	227	1.30	2.14	206	18	3
07869	315	1.35	2.32	281	29	5
07871	531	1.53	2.82	459	66	6
07874	94	1.37	1.92	87	7	0
07875	15	2.14	2.54	12	3	0
07876	138	1.13	1.46	132	6	0
07878	12	1.00	1.22	12	0	0
07879	7	1.82	2.25	6	1	0
07882	72	2.39	4.36	49	21	2
07885	107	1.38	2.26	97	9	1
07901	28	1.30	4.83	24	2	2
07920	122	1.70	2.54	95	27	0
07921	8	1.04	1.47	7	1	0
07922	30	0.98	1.32	29	1	0
07924	100	2.82	6.85	64	26	10
07926	41	1.65	3.14	31	9	1
08801	149	2.28	3.89	115	30	4
08802	50	2.18	3.66	38	11	1
08804	33	1.58	2.39	28	5	0
08805	10	0.82	1.01	10	0	0
08807	106	1.25	1.71	96	10	0
08808	4	2.64	3.12	2	2	0
08809	220	6.65	35.40	94	80	46
08812	11	0.57	0.61	11	0	0
08816	19	0.91	1.11	19	0	0
08817	10	0.86	1.03	10	0	0
08820	13	1.00	1.27	13	0	0
08822	107	2.84	9.63	68	30	9
08824	8	1.08	1.41	8	0	0

Zip	Total	Mean	Avg.	–>4	4–>20	20–>
07961	55	1.43	2.36	46	9	0
07974	25	1.07	2.15	23	1	1
07976	20	3.08	5.20	13	5	2
07977	9	4.94	11.43	5	1	3
07978	8	12.26	27.55	3	1	4
07979	13	2.69	5.14	8	4	1
07980	13	1.93	2.53	10	3	0
07981	130	1.19	1.95	116	13	1
08003	4	1.93	2.25	4	0	0
08055	6	0.89	1.04	6	0	0
08057	4	0.84	0.95	4	0	0
08502	22	2.26	3.92	17	4	1
08512	6	0.91	1.06	6	0	0
08520	9	1.04	1.81	8	1	0
08525	13	1.93	3.11	10	3	0
08530	14	1.07	2.02	12	2	0
08534	12	1.22	1.39	12	0	0
08536	6	1.44	1.46	6	0	0
08540	107	1.66	3.23	89	17	1
08542	4	3.25	4.15	3	1	0
08550	9	1.06	1.18	9	0	0
08551	16	2.17	3.32	12	4	0
08559	17	1.67	3.20	16	0	1
08560	11	1.68	3.86	10	0	1
08619	5	0.72	0.82	5	0	0
08628	11	2.99	13.09	5	4	2
08638	9	2.96	8.83	6	2	1
08648	28	2.17	3.89	21	6	1
08690	5	0.67	0.74	5	0	0
08691	5	0.81	0.90	5	0	0
08825	41	2.33	7.66	30	6	5
08826	89	2.46	7.34	66	14	9
08827	70	2.98	8.01	42	24	4
08829	69	2.94	5.18	42	25	2
08830	4	0.46	0.45	4	0	0
08833	187	2.90	5.47	121	58	8
08835	8	0.87	1.19	7	1	0
08836	12	1.01	1.34	12	0	0
08840	7	1.41	2.01	5	2	0
08846	13	0.73	1.14	12	1	0
08848	68	1.70	2.70	54	13	1
08850	6	1.81	2.17	5	1	0
08853	20	1.54	2.16	17	3	0
08854	21	1.31	1.78	19	2	0
08857	7	0.76	0.82	7	0	0
08858	19	2.85	4.30	12	7	0
08859	4	0.90	0.95	4	0	0
08865	85	2.58	6.04	65	16	4
08867	49	2.26	4.44	36	11	2
08869	9	0.76	1.57	8	1	0
08870	23	2.06	2.75	21	2	0
08873	32	1.45	1.75	32	0	0
08876	70	1.12	2.14	63	6	1
08880	6	0.99	1.43	6	0	0
08885	5	3.27	4.83	3	2	0
08886	11	3.49	4.16	7	4	0
08889	54	1.78	2.63	43	11	0
08901	4	1.35	1.55	4	0	0
08902	15	1.88	4.24	11	3	1
08904	11	1.11	1.47	10	1	0

08800	5	3.16	5.46	3	2	0	10304	4	1.29	1.54	4	0	0
10306	8	0.64	0.71	8	0	0	10708	6	0.89	1.01	6	0	0
10308	5	0.35	0.34	5	0	0	10710	4	0.80	0.90	4	0	0
10312	8	0.49	0.55	8	0	0	10901	37	1.13	1.39	37	0	0
10314	14	0.63	0.68	14	0	0	10912	9	1.63	2.40	8	1	0
10465	5	0.46	0.44	5	0	0	10918	9	1.29	1.75	8	1	0
10471	9	0.56	0.62	9	0	0	10920	4	0.50	0.52	4	0	0
10502	4	0.98	1.11	4	0	0	10921	6	1.86	3.34	3	3	0
10504	5	1.36	1.93	4	1	0	10922	4	2.11	2.34	3	1	0
10506	6	0.79	0.93	6	0	0	10923	4	0.51	0.53	4	0	0
10509	55	1.47	2.37	49	5	1	10924	4	1.34	1.46	4	0	0
10510	4	0.85	0.95	4	0	0	10925	8	1.21	1.37	8	0	0
10512	118	2.09	4.11	86	28	4	10930	6	1.32	1.57	6	0	0
10514	13	0.83	1.04	13	0	0	10950	34	1.01	1.60	31	3	0
10516	15	1.57	2.45	13	2	0	10952	8	1.00	1.25	8	0	0
10520	27	1.05	1.28	27	0	0	10954	6	0.70	0.81	6	0	0
10522	7	1.25	1.29	7	0	0	10956	22	0.82	0.98	22	0	0
10523	6	0.70	0.82	6	0	0	10960	4	0.81	0.87	4	0	0
10524	33	2.12	2.99	26	7	0	10970	16	.74	0.85	16	0	0
10532	4	0.93	0.95	4	0	0	10974	5	0.53	0.57	5	0	0
10533	7	1.18	1.55	6	1	0	10977	16	.93	1.13	16	0	0
10536	14	0.94	1.24	13	1	0	10980	10	1.14	2.13	9	1	0
10537	9	2.22	3.97	7	2	0	10984	6	0.80	1.28	5	1	0
10538	7	1.34	2.18	6	1	0	10987	11	2.21	2.85	10	1	0
10540	5	1.05	1.08	5	0	0	10990	32	1.41	1.92	30	2	0
10541	77	1.45	2.20	69	7	1	11030	6	1.07	1.40	6	0	0
10543	4	1.24	1.96	3	1	0	11229	4	0.54	0.62	4	0	0
10547	15	1.27	2.37	13	2	0	11234	4	0.71	0.71	4	0	0
10549	12	0.84	0.90	12	0	0	11374	4	0.46	0.42	4	0	0
10560	9	0.93	1.09	9	0	0	11725	4	0.80	0.86	4	0	0
10562	15	0.77	0.90	15	0	0	11733	5	0.62	0.65	5	0	0
10566	61	1.24	2.53	52	8	1	11743	4	0.63	0.70	4	0	0

Zip	Total	Mean	Avg.	–>4	4–>20	20–>	Zip	Total	Mean	Avg.	–>4	4–>20	20–>
10570	7	0.96	1.11	7	0	0	11746	6	0.65	0.72	6	0	0
10576	5	1.15	1.28	5	0	0	11763	4	0.41	0.38	4	0	0
10578	5	0.91	1.13	5	0	0	11793	4	0.63	0.63	4	0	0
10579	39	2.29	3.12	28	11	0	11803	5	0.61	0.75	5	0	0
10580	10	0.89	1.09	10	0	0	12010	4	1.53	2.60	2	2	0
10583	26	1.17	1.48	25	1	0	12054	4	0.90	0.95	4	0	0
10589	8	2.00	3.21	6	2	0	12061	4	1.75	1.91	4	0	0
10590	5	1.51	2.12	4	1	0	12065	4	1.05	1.25	4	0	0
10598	43	0.89	1.09	43	0	0	12180	5	1.80	3.18	4	1	0
10605	4	0.95	1.16	4	0	0	12186	6	1.35	2.53	4	2	0
10704	5	0.70	0.80	5	0	0	12302	5	0.92	0.98	5	0	0
10705	6	0.72	0.90	6	0	0	12309	6	0.53	0.53	6	0	0
10706	5	1.03	1.31	5	0	0	12401	16	1.02	1.31	16	0	0
12404	5	1.45	1.55	5	0	0	14845	6	2.09	3.43	4	2	0
12477	4	1.29	2.78	3	1	0	14850	10	1.23	1.67	9	1	0
12518	8	1.12	1.49	8	0	0	15005	6	8.64	34.84	3	1	2
12520	4	0.58	0.61	4	0	0	15010	4	8.35	10.82	0	3	1
12531	8	2.12	2.80	6	2	0	15017	8	1.07	1.70	7	1	0
12533	23	1.99	2.87	14	9	0	15024	7	1.80	2.28	6	1	0
12540	5	3.65	5.79	1	4	0	15044	12	2.53	4.77	8	3	1
12545	4	5.08	6.99	1	3	0	15090	8	1.79	3.16	6	2	0
12550	17	1.71	2.59	14	3	0	15101	7	2.76	3.61	4	3	0
12561	5	1.58	2.62	4	1	0	15102	5	0.94	1.02	5	0	0
12563	21	1.69	2.56	16	5	0	15108	11	3.00	3.71	8	3	0
12564	8	1.19	1.29	8	0	0	15129	4	1.84	3.78	3	1	0
12570	5	0.78	0.87	5	0	0	15143	8	3.26	7.19	5	2	1
12571	4	0.85	1.45	4	0	0	15146	11	1.79	1.92	11	0	0
12572	4	1.59	1.65	4	0	0	15147	10	1.91	3.58	7	3	0
12582	11	1.45	2.16	9	2	0	15206	9	0.81	1.03	9	0	0
12590	7	0.93	1.08	7	0	0	15207	5	1.88	2.36	4	1	0

12603	19	1.31	2.70	17	1	1	15212	4	1.22	2.34	3	1	0
12754	4	0.77	0.82	4	0	0	15213	14	1.06	1.52	13	1	0
12801	4	0.79	0.82	4	0	0	15214	5	1.27	2.13	4	1	0
13021	5	1.62	1.99	5	0	0	15215	9	2.10	3.03	7	2	0
13027	6	0.66	0.72	6	0	0	15216	9	0.91	1.19	9	0	0
13031	7	2.97	4.98	5	2	0	15217	22	1.11	1.51	20	2	0
13035	4	0.80	0.94	4	0	0	15218	7	1.60	2.54	6	1	0
13066	5	3.72	5.42	2	3	0	15220	7	2.28	5.84	5	2	0
13104	13	4.47	5.97	7	6	0	15221	7	2.57	3.33	5	2	0
13108	12	1.44	1.63	12	0	0	15227	10	1.23	1.43	9	1	0
13215	31	3.35	8.01	18	12	1	15228	11	1.08	1.76	10	1	0
13440	7	3.13	6.59	4	3	0	15232	9	1.12	1.41	9	0	0
13601	7	0.82	0.99	7	0	0	15234	6	1.94	3.08	4	2	0
13676	4	0.70	3.73	4	0	0	15235	14	3.07	4.75	7	7	0
13760	11	1.26	1.81	9	2	0	15236	9	2.17	2.61	7	2	0
13790	4	1.60	2.14	3	1	0	15237	24	2.16	3.28	16	8	0
13827	9	2.28	3.19	7	2	0	15238	5	4.09	4.59	2	3	0
13850	6	1.26	1.40	6	0	0	15239	7	2.05	2.84	6	1	0
13903	5	0.50	0.54	5	0	0	15241	18	2.37	3.47	11	7	0
13905	8	1.12	1.54	7	1	0	15243	13	4.01	9.21	9	3	1
14047	5	1.43	1.44	5	0	0	15301	8	1.18	1.56	7	1	0
14063	5	2.85	6.16	4	0	1	15317	15	1.53	2.84	14	0	1
14075	19	0.94	1.89	18	1	0	15601	8	1.37	2.44	6	2	0
14221	5	0.82	1.44	4	1	0	15642	7	1.40	2.63	5	2	0
14226	6	0.79	1.42	5	1	0	15650	5	2.50	5.44	3	2	0
14450	5	1.22	1.31	5	0	0	15656	6	2.06	2.30	5	1	0
14616	4	0.38	0.32	4	0	0	15668	8	2.74	5.15	5	3	0
15701	5	1.03	1.82	4	1	0	17403	12	7.73	17.64	4	3	5
15717	4	0.44	0.44	4	0	0	17404	9	1.71	2.25	7	2	0
15801	5	1.89	2.54	4	1	0	17501	8	0.01	14.85	2	3	3
15905	6	2.37	5.94	5	0	1	17502	8	3.75	4.09	4	4	0
16001	4	3.08	3.92	2	2	0	17508	8	22.20	26.40	0	3	5

Zip	Total	Mean	Avg.	->4	4->20	20->
16201	4	5.12	6.51	2	2	0
16222	5	2.94	3.72	2	3	0
16365	7	1.03	1.14	7	0	0
16509	4	1.11	1.86	3	1	0
16801	8	6.65	8.36	2	5	1
16803	7	8.80	13.17	2	3	2
16823	4	4.32	5.66	2	2	0
16925	4	3.25	4.01	3	1	0
17003	13	3.79	7.24	7	4	2
17011	61	6.46	10.67	17	36	8
17013	25	4.55	7.56	11	12	2
17019	8	2.35	2.69	7	1	0
17022	6	3.14	3.90	3	3	0
17025	7	2.41	2.73	6	1	0
17026	4	3.86	8.58	2	1	1
17033	18	3.67	7.78	9	8	1
17036	8	6.66	11.01	3	4	1
17038	7	11.60	19.37	2	2	3
17042	68	4.55	7.90	30	31	7
17043	4	5.39	8.03	2	2	0
17055	75	5.86	8.67	24	47	4
17067	7	4.01	4.66	5	2	0
17070	7	1.97	5.50	5	1	1
17073	7	1.80	2.78	5	2	0
17078	7	2.29	3.59	6	1	0
17109	5	5.12	7.22	2	3	0
17110	15	4.07	5.86	5	10	0
17111	7	7.54	16.57	3	1	3
17112	13	2.72	3.32	9	4	0
17201	26	3.09	4.72	17	8	1
17512	10	3.90	10.93	7	1	2
17517	7	2.42	2.94	4	3	0
17520	13	8.00	12.73	3	7	3
17522	32	4.56	11.44	14	15	3
17527	5	1.78	3.53	4	1	0
17532	11	6.09	9.43	3	6	2
17538	12	6.82	9.17	3	7	2
17540	13	5.74	10.56	4	7	2
17543	54	6.98	14.01	18	23	13
17545	11	5.97	9.85	4	5	2
17551	9	2.01	2.67	7	2	0
17555	8	2.08	2.95	7	1	0
17557	13	3.86	5.75	7	5	1
17560	4	2.86	6.20	2	2	0
17566	6	3.60	11.19	5	0	1
17578	9	2.79	4.59	6	3	0
17579	4	6.60	7.23	1	3	0
17582	5	2.47	3.47	3	2	0
17584	7	5.44	7.09	3	4	0
17601	85	5.45	8.90	31	47	7
17602	17	2.97	8.06	12	4	1
17603	57	2.95	4.29	33	24	0
17701	5	3.41	13.23	3	1	1
17752	4	86.92	97.73	0	0	4
17815	8	2.14	3.95	6	2	0
17837	6	1.94	4.83	4	2	0
17881	4	0.64	0.84	4	0	0
17960	5	2.49	3.20	4	1	0
17961	4	3.12	3.94	3	1	0
18001	7	0.90	1.04	7	0	0

17236	5	7.34	8.74	1	4	0
17257	9	2.75	4.57	5	4	0
17268	11	3.35	7.60	6	4	1
17302	4	6.20	9.97	1	2	1
17315	5	5.27	17.02	3	0	2
17339	6	2.77	3.31	4	2	0
17349	4	3.96	4.60	1	3	0
17363	4	11.73	17.12	1	1	2
17402	15	3.29	13.00	10	4	1
18037	4	3.89	5.63	3	1	0
18041	5	2.68	4.63	3	2	0
18042	277	3.67	18.72	161	83	33
18044	4	2.04	2.25	4	0	0
18049	56	3.22	5.46	32	21	3
18051	9	5.29	10.56	3	4	2
18052	15	3.69	10.11	9	4	2
18055	26	3.46	6.26	14	11	1
18056	4	7.56	12.63	2	0	2
18062	58	3.80	6.78	26	28	4
18064	16	3.72	9.51	11	3	2
18067	17	5.56	9.71	8	6	3
18069	20	7.17	18.93	7	4	9
18072	4	3.25	7.24	2	2	0
18077	12	3.30	4.54	6	6	0
18078	19	8.57	23.54	6	7	6
18086	4	9.58	22.17	1	1	2
18087	7	0.69	1.15	6	1	0
18088	8	6.64	8.51	2	6	0
18092	10	3.12	5.00	5	4	1
18102	48	1.65	3.06	40	8	0
18103	212	3.76	7.04	109	90	13
18104	214	3.92	6.62	103	101	10
18013	8	3.31	21.40	6	1	1
18014	9	4.39	7.96	6	1	2
18015	62	2.73	4.50	41	20	1
18017	208	4.18	6.46	97	100	11
18018	72	2.68	4.12	45	26	1
18031	5	3.83	17.14	3	1	1
18032	6	3.28	4.25	3	3	0
18034	30	7.45	23.82	10	11	9
18036	73	3.00	6.24	45	24	4
18914	7	3.32	6.70	4	2	1
18920	9	3.77	6.16	6	2	1
18931	4	1.46	1.55	4	0	0
18938	6	2.81	5.28	5	0	1
18940	12	2.76	5.49	8	3	1
18944	4	0.89	1.08	4	0	0
18947	4	1.75	2.13	4	0	0
18950	4	2.52	2.74	3	1	0
18951	9	1.62	2.72	7	2	0
18960	4	1.47	1.52	4	0	0
18966	15	2.89	5.24	8	6	1
18969	5	1.91	2.56	4	1	0
18972	5	1.32	1.78	4	1	0
18974	5	4.04	7.30	2	2	1
18977	5	2.35	2.48	5	0	0
19001	4	0.83	0.86	4	0	0
19002	12	1.79	2.55	10	2	0
19003	4	0.67	0.82	4	0	0
19004	12	2.71	3.48	8	4	0
19006	12	1.69	1.81	12	0	0
19010	14	1.46	4.50	11	1	2
19020	5	1.57	1.76	5	0	0
19025	4	2.18	2.51	4	0	0

Zip	Total	Mean	Avg.	->4	4->20	20->
18105	16	1.51	2.84	13	3	0
18106	18	3.20	3.97	10	8	0
18201	5	1.04	1.17	5	0	0
18210	6	1.41	2.01	6	0	0
18235	8	5.23	13.42	4	1	3
18301	10	1.42	1.75	9	1	0
18328	9	0.92	1.18	9	0	0
18335	4	0.89	1.25	4	0	0
18337	11	1.13	1.64	9	2	0
18353	7	5.47	17.03	4	1	2
18360	9	2.97	7.92	6	2	1
18411	4	1.26	1.32	4	0	0
18426	4	1.75	4.02	2	2	0
18428	15	1.34	1.63	14	1	0
18431	12	1.69	4.31	9	2	1
18436	6	1.17	1.29	6	0	0
18453	4	0.93	0.98	4	0	0
18657	5	2.00	2.29	4	1	0
18702	4	1.86	3.80	3	1	0
18707	4	1.52	1.62	4	0	0
18901	21	2.20	2.65	19	2	0
19320	4	2.76	3.37	3	1	0
19333	6	2.68	2.82	6	0	0
19335	28	1.78	2.50	23	5	0
19341	9	1.26	1.31	9	0	0
19343	8	1.04	1.40	8	0	0
19344	7	2.24	3.65	5	2	0
19348	5	1.27	1.37	5	0	0
19352	5	1.20	3.20	4	1	0
19355	10	2.30	2.92	7	3	0

Zip	Total	Mean	Avg.	->4	4->20	20->
19035	7	2.85	7.58	5	1	1
19038	5	2.27	4.16	4	1	0
19040	5	1.73	2.47	4	1	0
19041	4	2.52	3.12	3	1	0
19047	13	1.41	1.77	12	1	0
19063	11	2.44	2.97	9	2	0
19064	7	0.92	1.27	7	0	0
19066	8	1.62	2.42	6	2	0
19067	24	1.94	2.94	21	3	0
19072	5	2.70	4.56	4	1	0
19073	6	1.20	1.28	6	0	0
19081	6	0.92	1.12	6	0	0
19085	10	9.96	1.25	10	0	0
19087	28	3.67	9.83	14	11	3
19116	4	1.74	2.14	3	1	0
19118	5	1.75	2.24	4	1	0
19119	4	1.25	2.02	3	1	0
19151	7	0.63	0.71	7	0	0
19301	6	5.83	15.49	2	3	1
19312	5	1.51	1.71	5	0	0
19317	8	0.98	1.16	8	0	0
19606	43	6.08	9.87	12	26	5
19607	12	2.52	3.49	8	4	0
19610	7	4.55	6.42	2	5	0
20002	4	0.83	1.01	4	0	0
20003	13	0.79	0.99	13	0	0
20007	21	0.71	0.85	21	0	0
20008	15	1.22	1.72	14	1	0
20009	6	0.44	0.46	6	0	0
20011	9	1.02	1.15	9	0	0

19380	7	2.39	5.73	5	2	0	20012	7	1.21	1.56	6	1	0
19382	15	1.24	1.72	14	1	0	20015	22	0.95	1.35	20	2	0
19401	14	3.11	11.69	10	1	3	20016	27	1.34	1.74	25	2	0
19403	17	1.46	2.00	16	1	0	20017	6	1.33	1.97	5	1	0
19406	9	6.99	14.47	4	3	2	20018	9	0.72	0.74	9	0	0
19422	5	1.47	1.73	5	0	0	20020	4	0.94	1.72	3	1	0
19425	8	2.27	2.65	7	1	0	20024	4	1.01	1.11	4	0	0
19426	7	3.60	4.33	5	2	0	20036	6	0.30	0.16	6	0	0
19438	5	1.39	1.63	5	0	0	20601	10	1.00	1.17	10	0	0
19440	13	1.58	1.96	13	0	0	20617	4	0.91	1.02	4	0	0
19446	19	1.15	1.44	18	1	0	20639	7	2.55	3.25	3	4	0
19454	9	0.87	1.07	9	0	0	20646	6	1.76	2.44	5	1	0
19460	7	2.60	18.32	5	0	2	20678	4	1.30	1.53	4	0	0
19464	36	2.07	2.95	31	5	0	20705	11	0.89	1.05	11	0	0
19470	4	2.33	2.56	4	0	0	20706	19	1.36	1.63	17	2	0
19475	5	1.69	3.23	4	1	0	20707	36	1.97	3.33	29	6	1
19481	4	1.14	1.39	4	0	0	20708	6	2.07	8.16	5	0	1
19504	13	5.96	10.65	4	7	2	20715	27	1.95	3.28	23	3	1
19505	7	3.96	9.33	5	0	2	20716	7	1.06	1.45	7	0	0
19508	5	1.26	2.81	4	1	0	20735	9	1.64	2.01	8	1	0
19512	55	1.99	4.54	42	10	3	20737	4	1.31	2.09	3	1	0
19518	11	4.15	7.59	7	3	1	20740	13	0.85	1.07	13	0	0
19520	8	1.42	1.90	7	1	0	20742	4	0.50	0.50	4	0	0
19522	9	7.36	9.44	1	7	1	20743	6	2.20	5.05	5	0	1
19525	9	2.33	3.44	8	1	0	20744	16	1.35	1.87	15	1	0
19529	6	3.73	6.10	4	2	0	20745	6	2.13	2.94	5	1	0
19530	6	5.97	12.37	3	2	1	20746	5	2.28	3.39	3	2	0
19533	4	4.30	12.71	2	1	1	20748	15	1.25	2.88	14	0	1
19539	5	1.68	1.80	5	0	0	20754	9	3.15	4.79	6	3	0
19540	5	5.75	17.29	4	0	1	20759	4	6.39	6.45	0	4	0
19547	17	5.18	7.89	8	8	1	20770	11	0.66	0.75	11	0	0
19560	23	11.13	19.58	4	10	9	20772	9	2.03	3.11	6	3	0

Zip	Total	Mean	Avg.	–>4	4–>20	20–>
19603	12	1.36	2.44	9	3	0
19604	8	8.06	14.73	2	4	2
19605	8	4.69	7.51	4	3	1
20784	18	1.03	1.71	16	2	0
20814	41	1.18	2.12	36	4	1
20815	68	1.32	1.85	63	5	0
20816	38	2.04	3.00	31	7	0
20817	106	1.47	2.23	91	15	0
20818	5	4.65	5.53	3	2	0
20832	32	1.68	2.79	28	3	1
20833	7	1.78	1.97	7	0	0
20837	9	1.69	2.33	7	2	0
20841	4	2.64	4.31	2	2	0
20850	52	1.74	2.34	45	7	0
20851	9	2.02	2.56	7	2	0
20852	53	1.50	2.15	47	6	0
20853	63	1.33	1.69	61	2	0
20854	172	2.34	3.34	128	44	0
20855	47	1.91	2.55	38	9	0
20866	11	1.60	1.95	10	1	0
20871	22	3.19	4.57	14	7	1
20872	59	2.75	4.71	40	17	2
20874	40	1.99	3.49	33	7	0
20877	21	1.92	2.86	18	3	0
20878	92	2.21	3.06	73	18	1
20879	111	2.54	3.81	76	34	1
20895	39	1.86	2.28	34	5	0
20896	7	2.07	3.39	6	1	0
20901	33	2.13	2.97	27	6	0
20902	64	1.99	3.68	50	13	1
20777	5	2.40	2.49	5	0	0
20782	9	1.30	1.88	8	1	0
20783	16	1.41	1.93	14	2	0
21093	14	1.46	1.89	13	1	0
21114	10	1.25	1.57	10	0	0
21117	9	4.76	5.92	4	5	0
21120	4	8.02	12.32	1	2	1
21122	4	0.86	0.89	4	0	0
21133	5	1.18	1.62	4	1	0
21140	4	2.25	2.79	3	1	0
21144	4	0.62	0.64	4	0	0
21146	6	0.78	0.95	6	0	0
21157	15	9.18	18.65	3	9	3
21204	5	1.75	3.02	3	2	0
21208	17	1.28	2.12	14	3	0
21209	5	1.12	1.33	5	0	0
21210	4	1.14	1.84	3	1	0
21212	4	1.46	2.27	3	1	0
21227	5	0.77	0.82	5	0	0
21228	5	1.39	1.43	5	0	0
21234	6	1.64	5.02	5	0	1
21401	16	2.05	2.58	15	1	0
21403	7	0.83	1.09	7	0	0
21502	4	4.61	7.44	2	1	1
21550	4	2.54	3.42	2	2	0
21701	49	2.64	4.63	30	18	1
21738	6	4.33	5.30	3	3	0
21740	17	5.43	8.65	7	8	2
21754	22	3.89	8.40	11	8	3
21769	8	1.27	1.62	8	0	0

20903	16	2.52	4.10	13	3	0
20904	83	1.88	2.62	70	13	0
20906	59	1.25	1.54	58	1	0
20910	34	1.29	1.96	30	4	0
20912	20	0.66	0.78	20	0	0
21012	8	1.25	1.43	8	0	0
21029	9	4.15	5.48	4	5	0
21030	5	2.78	3.97	3	2	0
21032	7	1.29	1.91	6	1	0
21037	5	1.11	1.31	5	0	0
21043	53	3.15	4.40	27	25	1
21044	71	3.74	5.67	36	32	3
21045	58	3.11	4.64	34	24	0
21046	21	4.73	7.37	10	9	2
21047	5	5.87	11.07	3	1	1
21061	6	0.65	0.75	6	0	0
21076	4	0.79	0.97	4	0	0
22042	32	1.21	1.54	29	3	0
22043	19	1.50	1.85	18	1	0
22044	12	1.55	2.61	10	2	0
22046	10	0.93	1.19	9	1	0
22065	5	1.42	2.06	4	1	0
22066	32	2.73	3.52	22	10	0
22069	8	2.28	1.72	7	1	0
22070	33	1.70	6.32	27	5	1
22071	45	1.61	2.19	38	7	0
22075	13	1.43	1.91	12	1	0
22079	7	1.17	1.55	7	0	0
22080	4	0.76	0.84	4	0	0
22090	11	1.81	2.38	9	2	0
22091	46	2.09	2.77	33	13	0
22094	7	2.29	5.44	5	2	0

21770	11	8.11	13.25	2	7	2
21771	108	5.01	8.04	41	59	8
21783	4	12.09	12.44	0	4	0
21784	13	2.20	3.18	11	2	0
21793	6	2.90	3.10	5	1	0
21797	16	4.53	7.95	7	7	2
22003	89	1.61	2.27	76	13	0
22015	37	1.54	2.04	33	4	0
22020	36	1.88	2.51	29	7	0
22021	26	1.89	2.65	24	2	0
22024	13	2.19	2.35	12	1	0
22030	42	2.24	3.27	34	8	0
22031	24	2.30	3.79	21	3	0
22032	29	1.88	2.42	24	5	0
22033	21	1.58	5.72	17	1	3
22039	20	1.59	1.99	17	3	0
22041	9	106	1.42	9	0	0
22310	26	1.01	1.25	26	0	0
22311	5	1.33	1.57	5	0	0
22312	17	1.29	2.02	16	1	0
22314	10	0.74	0.96	10	0	0
22401	7	1.19	1.27	7	0	0
22554	5	1.84	2.28	4	1	0
22601	9	3.65	6.17	5	3	1
22801	4	1.41	1.88	3	1	0
22901	6	2.46	7.22	4	1	1
23113	5	1.38	1.61	5	0	0
23229	4	1.51	1.59	4	0	0
23233	5	1.57	1.86	4	1	0
23235	8	2.04	2.57	6	2	0
23692	4	0.78	0.90	4	0	0
24018	12	2.68	3.45	8	4	0

Zip	Total	Mean	Avg.	–>4	4–>20	20–>	Zip	Total	Mean	Avg.	–>4	4–>20	20–>
22101	74	1.89	2.82	60	13	1	24060	4	2.53	4.02	3	1	0
22102	23	1.86	2.76	16	7	0	24450	7	1.79	1.95	7	0	0
22110	24	1.19	1.73	23	1	0	24502	7	3.16	4.96	4	3	0
22111	15	1.83	2.07	15	0	0	24540	6	1.76	2.01	6	0	0
22123	12	1.16	1.43	11	1	0	24541	6	2.82	2.98	5	1	0
22124	25	2.52	4.36	20	3	2	25411	4	3.53	12.64	2	0	2
22132	6	1.48	2.16	5	1	0	25425	25	1.85	5.47	19	3	3
22150	10	1.19	1.78	9	1	0	25443	6	2.46	3.18	5	1	0
22151	27	1.51	1.97	25	2	0	26003	6	1.98	6.45	5	0	1
22152	34	1.89	2.30	30	4	0	26505	5	1.28	1.47	5	0	0
22153	33	1.54	2.03	29	4	0	27106	6	5.95	10.65	3	2	1
22170	24	1.85	2.62	18	6	0	27408	4	1.63	2.29	3	1	0
22180	71	1.59	2.09	61	10	0	27410	5	1.45	1.57	5	0	0
22186	13	1.40	2.35	11	2	0	27511	5	1.87	2.14	5	0	0
22191	4	1.73	1.90	4	0	0	27604	5	2.61	2.83	4	1	0
22192	9	1.54	1.91	8	1	0	27606	5	1.05	1.31	5	0	0
22193	6	1.47	1.80	5	1	0	27607	7	0.87	1.10	7	0	0
22201	9	1.05	1.23	9	0	0	27612	6	1.23	1.88	5	1	0
22202	8	1.19	1.42	8	0	0	27705	4	1.72	1.94	4	0	0
22203	6	1.40	1.80	6	0	0	28021	14	3.29	4.47	9	5	0
22204	9	1.50	1.81	9	0	0	28110	4	1.13	1.36	4	0	0
22205	7	1.46	1.99	6	1	0	28601	4	1.69	1.79	4	0	0
22207	38	1.13	1.68	35	3	0	28711	4	1.24	1.35	4	0	0
22301	4	0.79	0.96	4	0	0	28734	21	2.25	2.67	18	3	0
22302	14	1.24	1.41	14	0	0	28739	37	4.60	7.09	15	20	2
22304	5	1.34	1.51	5	0	0	28779	6	1.08	1.24	6	0	0
22306	8	0.59	0.71	8	0	0	28804	4	4.13	5.45	2	2	0
22308	23	0.98	1.20	23	0	0	29418	5	0.65	0.70	5	0	0
22309	15	1.26	1.60	14	1	0	30062	4	2.19	2.24	4	0	0
30075	7	1.22	1.38	1	0	0	37203	30	1.82	2.96	21	9	0
30087	7	3.38	3.68	5	2	0	37204	9	5.09	7.07	4	4	1

30327	9	1.94	2.48	8	1	0
32073	4	0.58	0.61	4	0	0
32201	17	0.59	0.62	17	0	0
32206	10	1.17	1.40	9	1	0
32548	5	1.27	1.34	5	0	0
32578	7	0.69	0.75	7	0	0
32630	7	1.80	1.92	7	0	0
33507	4	0.85	0.94	4	0	0
33511	7	1.86	2.20	7	0	0
33526	4	1.30	1.59	4	0	0
33529	4	1.37	1.70	4	0	0
33533	5	0.63	0.68	5	0	0
33542	4	2.21	3.15	3	1	0
33566	4	1.64	2.23	4	0	0
33570	11	1.48	1.74	10	1	0
33582	6	1.49	1.76	5	1	0
33583	10	1.06	1.35	10	0	0
33584	7	2.87	3.85	5	2	0
33595	12	0.85	1.03	12	0	0
33615	5	1.39	1.53	5	0	0
33617	5	1.18	1.54	5	0	0
33618	4	0.91	1.10	4	0	0
33624	7	1.10	1.12	7	0	0
33629	4	0.50	0.52	4	0	0
33803	43	2.74	5.00	25	17	1
33830	7	6.51	17.72	3	3	1
33860	7	2.71	3.65	6	1	0
33880	5	1.47	1.78	5	0	0
34289	6	1.86	3.52	3	3	0
35801	4	2.05	2.25	3	1	0
35803	5	3.46	6.13	3	2	0
37013	11	1.85	2.51	10	1	0
37014	6	2.84	4.72	4	2	0
37205	38	2.61	3.21	30	8	0
37207	5	2.93	3.21	4	1	0
37208	8	1.82	3.79	5	3	0
37209	10	2.94	3.62	6	4	0
37211	9	3.89	6.38	6	2	1
37212	14	1.79	2.93	12	2	0
37214	5	2.22	2.82	4	1	0
37215	36	3.10	4.54	18	17	1
37216	4	9.70	10.19	0	4	0
37217	6	0.93	1.10	6	0	0
37220	6	3.46	3.96	4	2	0
37221	6	1.36	1.55	6	0	0
37311	6	1.48	1.73	6	0	0
37355	8	1.40	1.91	7	1	0
37363	6	1.92	2.00	6	0	0
37377	8	1.15	1.24	8	0	0
37388	4	1.82	3.04	3	1	0
37415	6	1.36	1.81	5	1	0
37421	6	1.47	1.87	6	0	0
37601	7	4.49	6.02	3	4	0
37659	5	2.26	2.77	4	1	0
37664	4	3.63	4.50	2	2	0
37763	5	1.88	3.04	3	2	0
37801	6	3.41	3.74	4	2	0
37814	6	1.62	2.06	5	1	0
37830	18	3.01	4.29	11	7	0
37920	11	1.15	1.78	9	2	0
37922	7	3.56	4.38	3	4	0
37923	10	2.91	3.62	9	1	0
38501	5	2.69	4.61	3	2	0
38555	4	0.77	0.96	4	0	0
38553	6	4.01	14.63	3	2	1
40205	4	3.58	3.81	3	1	0

Zip	Total	Mean	Avg.	–>4	4–>20	20–>	Zip	Total	Mean	Avg.	–>4	4–>20	20–>
37027	183	2.51	3.65	132	48	3	40207	6	3.33	3.87	5	1	0
37055	4	1.73	1.81	4	0	0	40222	4	3.61	4.45	1	3	0
37064	131	2.54	3.58	96	33	2	40223	4	3.29	5.15	2	2	0
37075	17	1.91	2.46	15	2	0	40403	6	2.85	3.58	4	2	0
37076	6	1.57	1.82	6	0	0	40502	10	5.85	9.09	2	7	1
37115	5	1.03	1.16	5	0	0	40503	4	2.25	7.50	1	3	0
37122	8	1.55	1.71	8	0	0	40505	6	3.39	5.39	4	2	0
37130	11	2.20	2.57	10	1	0	42101	4	1.12	2.13	3	1	0
37172	4	1.34	1.60	4	0	0	43081	8	3.01	3.60	4	4	0
43209	4	3.51	5.74	3	1	0	47803	6	2.46	2.99	5	1	0
43213	4	5.25	5.99	1	3	0	47906	4	1.91	2.08	4	0	0
43214	8	5.52	7.68	3	4	1	48010	6	1.36	1.88	6	0	0
43220	6	5.26	6.33	2	4	0	48013	6	1.15	1.53	6	0	0
43221	11	6.36	7.99	2	8	1	48018	6	1.56	1.80	6	0	0
43230	5	3.45	9.14	3	1	1	48024	4	0.86	0.94	4	0	0
44022	5	1.41	2.03	4	1	0	48033	6	1.73	2.45	5	1	0
44039	4	2.89	3.29	3	1	0	48098	4	1.26	1.54	4	0	0
44060	4	1.82	2.62	3	1	0	48103	13	2.06	2.46	11	2	0
44116	5	1.32	1.60	5	0	0	48104	6	1.79	2.45	6	0	0
44122	5	0.71	0.81	5	0	0	48105	8	1.22	1.35	8	0	0
44124	7	1.01	1.16	7	0	0	48127	4	1.40	1.48	4	0	0
44130	4	1.24	1.27	4	0	0	48154	5	0.85	0.85	5	0	0
44136	8	1.60	1.91	8	0	0	48161	4	2.04	2.31	3	1	0
44256	7	3.10	3.70	4	3	0	48167	16	1.67	2.08	15	1	0
44312	5	3.63	4.12	2	3	0	48189	4	3.92	4.46	3	1	0
44313	4	2.04	2.17	4	0	0	48230	4	1.26	1.28	4	0	0
44720	4	2.98	4.23	3	1	0	48236	5	0.73	0.78	5	0	0
45342	6	1.65	1.87	6	0	0	48439	4	0.86	1.05	4	0	0
45373	11	3.31	4.98	7	4	0	48823	4	1.44	1.59	4	0	0
45383	7	2.27	3.12	4	3	0	48864	5	0.96	1.40	4	1	0
45419	4	2.40	2.58	3	1	0	49001	7	3.09	3.91	4	3	0

45420	5	2.80	2.95	4	1	0	49007	6	1.36	1.84	5	1	0
45424	6	2.24	3.35	4	2	0	49010	4	0.75	0.76	4	0	0
45429	6	4.06	4.96	4	2	0	49017	4	0.84	0.98	4	0	0
45432	4	3.03	4.38	3	1	0	49068	4	1.82	2.50	3	1	0
45440	5	2.29	2.81	4	1	0	49127	4	1.12	1.35	4	0	0
45459	16	2.95	4.91	11	4	1	49301	5	2.08	4.37	3	2	0
46011	4	3.91	4.42	2	2	0	49341	4	1.08	1.38	4	0	0
46206	5	6.46	15.42	2	0	3	49418	4	3.64	5.70	3	1	0
46220	14	4.82	5.86	5	9	0	49428	5	3.22	4.10	4	1	0
46321	5	1.23	1.53	5	0	0	49450	7	1.94	2.03	7	0	0
46341	4	2.15	3.70	3	1	0	49456	4	0.97	1.06	4	0	0
46342	5	1.05	1.48	5	0	0	49504	8	1.12	1.33	8	0	0
46350	6	2.89	3.70	4	2	0	49505	18	1.06	1.14	18	0	0
46383	5	1.98	2.36	5	0	0	49506	25	1.06	1.27	25	0	0
46923	4	13.36	26.81	1	1	2	49507	5	1.24	1.41	5	0	0
47126	18	1.66	2.40	16	2	0	49508	10	0.75	0.91	10	0	0
47130	17	2.15	2.85	12	5	0	49509	7	1.09	1.90	6	1	0
47143	9	3.77	7.62	5	3	1	49855	4	0.83	0.90	4	0	0
47150	23	1.76	2.17	19	4	0	50010	5	3.89	5.34	2	3	0
47172	18	3.31	8.79	11	5	2	51360	5	1.83	2.44	4	1	0
47401	5	1.76	2.21	4	1	0	52101	4	2.77	3.51	3	1	0
47802	14	2.14	2.76	13	1	0	53092	8	2.06	2.48	8	0	0
53211	6	1.16	1.31	6	0	0	60098	7	2.85	4.95	4	3	0
54143	4	0.52	0.51	4	0	0	60101	4	1.13	1.35	4	0	0
54401	6	5.31	9.20	2	3	1	60103	6	2.02	2.27	6	0	0
54449	6	19.95	20.97	0	2	4	60106	7	1.92	3.10	6	1	0
54911	4	1.49	1.70	4	0	0	60108	12	1.20	1.36	12	0	0
54981	5	3.06	4.78	3	2	0	60126	38	1.75	2.84	31	6	1
55119	4	1.02	1.58	3	1	0	60134	5	1.99	2.37	4	1	0
55344	7	1.30	1.42	7	0	0	60135	4	2.10	2.28	4	0	0
56001	6	3.01	3.91	4	2	0	60137	18	2.06	2.63	14	4	0
56301	4	2.51	2.80	3	1	0	60143	5	2.44	3.89	3	2	0
56345	4	0.82	1.02	4	0	0	60148	14	1.56	2.15	11	3	0

Zip	Total	Mean	Avg.	–>4	4–>20	20–>
56560	7	1.48	2.43	6	1	0
58103	13	2.67	4.83	7	6	0
60004	59	1.42	1.70	56	3	0
60005	28	1.55	1.81	27	1	0
60007	9	1.09	1.22	9	0	0
60008	7	2.23	3.04	5	2	0
60010	21	1.59	1.95	17	4	0
60014	10	1.62	2.14	8	2	0
60015	19	1.52	2.01	17	2	0
60016	15	1.46	2.80	11	4	0
60022	11	0.97	1.22	11	0	0
60025	22	1.79	2.18	21	1	0
60031	4	0.78	1.18	4	0	0
60035	22	1.36	1.75	20	2	0
60044	4	1.27	1.32	4	0	0
60045	18	1.10	1.28	18	0	0
60046	5	0.65	0.69	5	0	0
60047	15	1.70	2.22	13	2	0
60048	11	1.54	2.22	9	2	0
60050	7	2.64	2.82	5	2	0
60053	19	0.94	1.15	19	0	0
60056	21	1.45	1.90	19	2	0
60061	9	1.00	1.20	9	0	0
60062	19	1.55	1.77	18	1	0
60067	37	1.54	1.88	34	3	0
60068	12	1.27	1.69	11	1	0
60070	7	2.96	5.61	6	0	1
60076	7	0.62	0.70	7	0	0
60085	5	1.30	1.90	4	1	0
60089	12	1.15	1.61	12	0	0
60090	8	0.91	1.27	8	0	0
60153	6	1.21	1.32	6	0	0
60172	6	2.27	3.07	5	1	0
60174	10	3.82	4.86	6	4	0
60181	9	1.88	2.17	8	1	0
60185	8	2.92	3.25	6	2	0
60187	16	2.02	2.96	13	3	0
60188	8	1.94	2.43	5	3	0
60193	24	1.28	1.97	20	4	0
60194	8	1.08	1.24	8	0	0
60195	10	1.10	1.28	10	0	0
60201	9	0.54	0.57	9	0	0
60202	4	1.08	1.67	3	1	0
60305	5	1.22	1.32	5	0	0
60402	7	0.85	0.98	7	0	0
60406	4	1.92	2.32	3	1	0
60422	5	1.33	1.63	5	0	0
60430	11	1.29	1.84	10	1	0
60435	12	2.21	2.83	8	4	0
60439	8	1.88	2.88	5	3	0
60441	12	1.30	1.46	12	0	0
60450	4	1.13	1.27	4	0	0
60451	4	1.61	1.71	4	0	0
60452	6	1.09	1.40	6	0	0
60453	10	0.87	1.02	10	0	0
60462	10	1.98	2.61	8	2	0
60463	4	2.25	3.50	2	2	0
60464	9	2.56	4.09	7	2	0
60473	8	1.03	1.51	7	1	0
60477	6	0.77	0.89	6	0	0
60506	4	2.82	2.99	3	1	0
60510	15	2.18	2.78	11	4	0

60091	19	1.20	1.86	15	4	0	60514	8	1.54	1.92	8	0	0
60093	11	0.75	1.07	10	1	0	60515	8	2.74	4.00	3	5	0
60516	9	2.97	4.03	5	4	0	66208	5	3.94	4.18	2	3	0
60517	5	0.67	0.77	5	0	0	66212	5	1.15	1.22	5	0	0
60521	70	2.44	3.76	50	20	0	66214	5	4.44	6.30	2	3	0
60525	21	1.87	2.20	18	3	0	72653	4	1.20	1.39	4	0	0
60532	9	2.11	2.60	7	2	0	74009	4	0.99	1.85	3	1	0
60540	31	1.99	2.99	23	8	0	75229	4	3.35	3.61	3	1	0
60543	6	2.48	2.73	5	1	0	76109	4	6.41	7.19	1	3	0
60544	4	4.70	5.73	1	3	0	80020	4	3.29	3.39	3	1	0
60546	5	0.96	1.08	5	0	0	80112	5	3.05	3.47	3	2	0
60558	18	1.87	2.58	15	3	0	80122	4	4.32	4.84	2	2	0
60559	5	2.30	3.40	3	2	0	80127	6	3.04	3.79	4	2	0
60565	21	2.15	3.31	17	4	0	80132	10	3.16	3.35	7	3	0
60614	6	0.64	0.76	6	0	0	80220	12	7.36	10.90	3	7	2
60622	4	0.81	1.18	4	0	0	80222	5	2.99	6.50	2	3	0
60626	4	0.66	0.68	4	0	0	80226	6	3.77	5.42	3	3	0
60629	4	0.78	1.46	3	1	0	80301	9	5.66	8.78	4	4	1
60635	4	0.69	0.94	4	0	0	80302	30	3.22	4.67	20	8	2
60638	4	0.76	0.81	4	0	0	80303	16	4.37	6.16	6	10	0
60645	5	0.80	1.05	5	0	0	80401	10	2.43	3.64	8	2	0
60646	8	1.24	1.64	8	0	0	80403	21	2.32	4.18	16	4	1
60648	4	1.85	1.92	4	0	0	80433	8	7.90	25.98	2	5	1
60654	194	1.91	2.91	152	40	2	80439	19	3.77	4.62	9	10	0
60656	7	0.96	1.08	7	0	0	80465	4	3.46	3.47	4	0	0
61832	5	1.04	1.10	5	0	0	80466	5	1.88	1.94	5	0	0
62025	4	1.11	1.22	4	0	0	80501	17	2.84	3.29	12	5	0
62650	6	2.93	5.97	3	2	1	80524	6	4.95	5.29	2	4	0
62704	4	3.05	4.90	2	2	0	80525	4	2.31	6.10	3	1	0
63017	7	1.07	1.41	7	0	0	80537	5	2.18	2.79	4	1	0
63031	5	1.57	1.75	5	0	0	80863	15	5.10	5.84	6	9	0
63033	4	1.65	1.86	4	0	0	80904	10	5.27	6.96	3	6	1
63042	5	0.85	1.15	5	0	0	80906	24	7.46	11.11	6	14	4

Zip	Total	Mean	Avg.	–>4	4–>20	20–>
63122	5	1.49	2.07	4	1	0
63131	7	1.45	1.57	7	0	0
63601	4	1.03	1.09	4	0	0
63628	4	2.19	2.70	3	1	0
64055	5	1.12	1.93	4	1	0
64063	4	1.76	2.29	3	1	0
64111	9	2.42	7.97	7	1	1
64152	5	1.37	2.16	3	2	0
64801	16	0.82	1.04	16	0	0
66044	4	3.41	5.22	2	2	0
66202	4	2.27	2.38	4	0	0
66204	6	6.66	8.23	2	4	0
66206	5	3.78	4.61	3	2	0
83201	11	1.41	1.78	10	1	0
83204	4	2.11	2.31	3	1	0
83263	5	0.73	0.74	5	0	0
83276	9	1.64	2.50	7	2	0
83301	9	2.13	2.61	8	1	0
83318	7	1.09	2.29	6	1	0
83330	4	1.92	2.05	4	0	0
83333	28	4.24	7.61	13	14	1
83338	9	1.51	1.99	8	1	0
83340	11	4.36	5.39	5	6	0
83341	4	2.76	3.19	2	2	0
83350	6	2.89	3.42	5	1	0
83352	4	2.97	4.13	2	2	0
83353	11	4.13	5.38	5	6	0
83401	15	1.77	2.04	15	0	0
83402	9	1.85	2.20	8	1	0
83420	4	2.38	2.51	4	0	0
83467	6	1.19	2.20	4	2	0

Zip	Total	Mean	Avg.	–>4	4–>20	20–>
80907	19	2.09	2.56	17	2	0
80908	14	5.96	7.69	4	10	0
80909	7	2.51	2.56	7	0	0
80910	4	4.75	7.44	2	2	0
80917	13	2.38	3.18	10	3	0
80918	20	1.71	1.95	19	1	0
80919	7	1.95	2.12	7	0	0
81211	4	3.48	3.91	2	2	0
82331	4	3.70	10.65	3	0	1
82414	4	0.90	1.18	4	0	0
82520	22	4.89	7.36	11	10	1
82801	22	2.54	3.48	14	8	0
83001	14	2.25	4.02	12	2	0
83703	7	2.70	3.08	4	3	0
83704	16	2.83	3.06	13	3	0
83705	14	1.97	2.33	13	1	0
83706	15	2.51	2.81	13	2	0
83709	11	1.73	2.08	10	1	0
83712	14	1.88	2.34	12	2	0
83714	5	0.90	2.96	4	1	0
83720	4	0.95	1.30	4	0	0
83801	5	2.31	2.61	4	1	0
83805	11	1.40	1.83	10	1	0
83814	42	3.59	4.56	23	19	0
83835	18	3.10	6.13	10	7	1
83837	4	1.22	1.56	4	0	0
83843	19	1.90	2.65	15	4	0
83854	18	3.56	4.52	11	7	0
83858	8	5.89	21.92	4	2	2
83861	9	1.54	2.33	8	1	0
83864	15	1.50	1.88	14	1	0

83501	35	1.77	2.67	32	2	1
83530	16	1.65	3.10	12	4	0
83536	7	2.51	2.80	5	2	0
83543	4	1.25	1.49	4	0	0
83544	13	1.68	2.23	10	3	0
83604	4	1.13	1.15	4	0	0
83605	36	1.23	1.62	35	1	0
83610	6	1.82	1.89	6	0	0
83611	24	1.13	1.52	23	1	0
83612	8	1.82	2.45	6	2	0
83615	5	1.57	2.30	4	1	0
83616	9	1.78	2.51	7	2	0
83617	10	1.85	1.95	10	0	0
83629	4	1.51	2.63	3	1	0
83631	4	1.67	2.32	4	0	0
83638	16	0.81	0.91	16	0	0
83639	4	1.35	1.69	4	0	0
83641	4	3.92	4.79	2	2	0
83642	7	2.00	2.91	5	2	0
83647	16	5.16	8.50	7	8	1
83651	16	1.54	1.91	16	0	0
83654	4	1.45	1.63	4	0	0
83655	4	2.32	3.17	3	1	0
83661	5	2.01	2.58	4	1	0
83672	5	2.12	2.14	5	0	0
83702	48	2.40	3.00	35	13	0
98862	7	3.39	5.28	5	1	1
99114	4	4.07	7.15	2	1	1
99163	6	2.81	3.62	4	2	0
99203	8	4.11	9.89	4	3	1
99204	8	2.02	2.41	7	1	0
99206	8	4.99	8.77	4	2	2
99208	15	3.31	8.80	11	3	1
83869	4	22.82	63.65	1	1	2
84106	4	2.97	3.55	3	1	0
86001	6	2.27	3.61	4	2	0
86301	11	2.21	2.78	8	3	0
87111	4	5.74	9.20	1	2	1
87185	4	8.33	62.74	2	0	2
87501	4	3.52	4.07	2	2	0
87801	5	2.20	2.56	4	1	0
90049	4	1.26	1.41	4	0	0
90067	4	0.55	0.69	4	0	0
91105	4	0.45	0.46	4	0	0
91436	5	1.27	1.46	5	0	0
92025	6	0.80	0.83	6	0	0
92376	4	0.54	0.57	4	0	0
92397	4	0.53	0.57	4	0	0
92631	4	1.71	1.84	4	0	0
95052	6	1.47	1.55	6	0	0
95405	5	0.74	0.80	5	0	0
95472	5	1.52	9.78	4	0	1
95730	5	2.81	4.91	4	1	0
96130	5	2.08	2.72	3	2	0
97405	6	0.43	0.38	6	0	0
97520	11	2.89	6.30	7	3	1
98248	13	0.30	0.12	13	0	0
98498	4	1.64	2.00	4	0	0
98801	9	1.11	1.21	9	0	0
99212	4	4.94	8.00	1	3	0
99216	9	13.81	17.21	0	6	3
99218	4	4.06	5.03	1	3	0
99223	6	3.35	5.20	3	3	0
99352	5	2.01	2.06	5	0	0
99516	5	1.02	1.58	5	0	0
99827	4	1.15	1.18	4	0	0

TABLE 3–C. ZIP CODES BY 5 DIGITS FOR BASEMENTS (ONLY GROUPS WITH COUNTS > = 4 ARE PRINTED.)

Zip	Total	Mean	Avg.	–>4	4–>20	20–>	Zip	Total	Mean	Avg.	–>4	4–>20	20–>
00000	89	1.97	4.19	64	23	2	07042	13	1.37	1.51	13	0	0
01451	5	10.22	12.22	0	4	1	07043	11	1.79	1.96	11	0	0
01741	4	2.79	4.53	2	2	0	07044	7	1.24	1.43	7	0	0
01742	6	3.74	4.59	4	2	0	07045	28	1.74	2.72	22	6	0
01760	6	1.42	3.11	5	1	0	07046	38	1.85	2.25	35	3	0
01775	5	3.85	5.32	1	4	0	07052	8	1.21	1.32	8	0	0
01824	7	1.79	2.61	6	1	0	07054	56	1.40	1.94	51	5	0
01886	6	10.36	12.19	1	4	1	07055	8	1.74	1.83	8	0	0
02026	4	3.83	3.98	2	2	0	07058	8	2.17	3.06	5	3	0
02114	7	2.16	3.19	5	2	0	07060	28	2.67	6.04	20	6	2
02174	6	2.65	2.71	6	0	0	07063	5	1.64	1.76	5	0	0
02181	10	1.08	1.76	9	1	0	07068	4	1.60	2.01	4	0	0
02192	6	2.61	4.16	4	2	0	07070	5	1.98	2.35	4	1	0
02194	4	2.56	3.12	3	1	0	07078	11	1.57	1.98	10	1	0
03449	4	1.67	2.23	4	0	0	07079	6	1.03	1.35	6	0	0
06066	5	2.57	4.51	4	1	0	07081	5	0.75	0.79	5	0	0
06107	4	3.22	6.90	2	1	1	07082	21	3.71	9.44	13	6	2
06484	4	1.86	1.99	4	0	0	07090	7	1.96	2.07	7	0	0
06525	6	5.45	7.27	2	4	0	07092	4	3.22	6.45	2	2	0
06812	5	1.29	2.01	4	1	0	07110	7	1.81	2.06	6	1	0
06840	4	1.75	2.08	4	0	0	07401	6	1.31	1.48	6	0	0
06877	4	4.06	5.67	2	2	0	07403	16	1.55	1.99	15	1	0
06880	5	7.99	11.77	1	3	1	07405	42	2.11	2.96	34	8	0
07003	9	0.89	1.29	9	0	0	07410	13	0.88	0.95	13	0	0
07005	42	2.19	3.12	31	11	0	07416	5	2.68	3.39	3	2	0
07006	12	1.57	1.95	10	2	0	07417	4	3.82	5.32	3	1	0

07009	6	1.80	3.75	5	1	0	07418	4	2.12	2.65	3	1	0
07013	7	1.57	1.72	7	0	0	07419	7	4.55	12.16	4	0	3
07024	4	1.10	1.35	4	0	0	07420	4	2.68	2.70	4	0	0
07028	4	2.68	3.24	3	1	0	07421	15	5.97	11.43	7	4	4
07032	6	1.38	1.41	6	0	0	07422	34	2.69	4.23	23	10	1
07034	7	1.72	2.01	7	0	0	07423	5	1.26	1.42	5	0	0
07039	18	1.81	3.12	17	0	1	07424	4	1.28	2.03	3	1	0
07040	10	1.58	1.73	10	0	0	07430	30	2.10	2.93	23	7	0
07432	4	1.04	1.40	4	0	0	07850	20	3.60	6.25	10	9	1
07435	7	5.22	10.85	4	2	1	07853	92	6.57	14.31	33	45	14
07436	26	1.46	1.87	23	3	0	07857	4	0.79	0.86	4	0	0
07438	40	2.76	4.54	24	16	0	07860	36	4.55	10.08	18	14	4
07439	6	3.37	6.16	4	1	1	07863	4	2.13	2.54	3	1	0
07440	4	0.99	1.06	4	0	0	07865	5	4.87	5.45	1	4	0
07442	6	4.10	4.68	4	2	0	07866	78	2.83	5.09	55	22	1
07444	16	2.03	2.54	14	2	0	07869	124	3.08	6.21	73	47	4
07446	11	1.37	1.86	10	1	0	07871	140	3.32	8.37	82	52	6
07450	13	1.54	2.46	11	2	0	07874	20	3.00	4.52	13	7	0
07452	5	1.28	1.43	5	0	0	07876	47	2.30	2.77	38	9	0
07456	55	1.59	1.98	52	3	0	07882	22	5.02	9.85	9	11	2
07458	15	1.30	1.63	15	0	0	07885	31	3.12	4.45	19	11	1
07460	13	1.85	2.81	10	3	0	07901	16	1.92	7.67	14	0	2
07461	40	3.75	5.91	19	18	3	07920	53	4.90	9.84	22	25	6
07462	26	4.38	6.64	9	15	2	07922	7	2.30	3.55	5	2	0
07465	10	1.48	1.75	9	1	0	07924	27	5.72	14.14	12	10	5
07470	41	1.86	2.27	36	5	0	07926	10	3.88	4.65	4	6	0
07480	44	2.42	3.58	30	13	1	07927	9	1.16	1.61	9	0	0
07481	11	1.33	1.74	11	0	0	07928	53	3.05	4.85	30	21	2
07506	4	0.93	1.25	4	0	0	07930	71	4.21	7.07	33	34	4
07603	22	1.43	1.82	20	2	0	07931	14	4.27	7.54	10	2	2
07628	5	1.39	1.58	5	0	0	07932	14	1.11	1.67	13	1	0
07631	5	1.93	2.20	4	1	0	07933	4	1.57	1.93	4	0	0

Zip	Total	Mean	Avg.	->4	4->20	20->
07642	4	1.70	1.89	4	0	0
07646	4	1.15	1.45	4	0	0
07666	11	1.03	1.20	11	0	0
07670	5	1.78	2.31	4	1	0
07675	4	1.75	2.14	4	0	0
07801	46	3.57	6.65	28	15	3
07821	28	3.00	7.09	17	10	1
07823	6	8.46	9.12	0	6	0
07825	16	4.43	7.61	6	8	2
07826	4	7.01	8.69	1	3	0
07828	22	3.53	4.13	13	9	0
07830	48	6.02	9.75	17	28	3
07832	4	4.92	6.88	1	3	0
07834	82	2.19	2.68	61	21	0
07836	33	3.41	5.44	19	12	2
07840	37	3.92	6.75	16	18	3
07843	27	4.01	5.30	12	15	0
07844	5	7.57	10.86	2	2	1
07848	8	2.29	5.69	6	1	1
07849	15	2.75	3.44	9	6	0
08648	14	5.71	18.58	6	4	4
08801	54	6.29	16.73	20	27	7
08802	24	5.63	10.35	10	9	5
08804	18	3.54	5.14	8	10	0
08807	43	2.72	3.65	32	11	0
08809	53	15.35	48.72	10	20	23
08812	8	2.60	4.03	6	2	0
08816	5	1.59	2.42	4	1	0
08817	4	2.61	4.26	3	1	0
07934	14	14.28	34.54	4	3	7
07936	7	2.08	2.38	6	1	0
07940	16	2.73	5.54	11	4	1
07945	91	4.58	11.63	43	40	8
07946	7	2.82	3.91	6	1	0
07950	63	3.41	5.20	31	31	1
07960	141	2.93	7.05	88	44	9
07961	19	2.43	3.35	12	7	0
07974	9	1.47	1.77	9	0	0
07976	12	5.99	9.59	3	8	1
07978	4	8.21	15.0	4	2	0
07980	7	3.58	4.21	2	5	0
07981	25	1.58	2.21	21	4	0
08502	14	4.11	5.06	6	8	0
08525	7	4.11	6.92	4	2	1
08530	4	1.76	2.01	4	0	0
08540	57	4.55	8.70	29	24	4
08550	5	2.15	3.30	4	1	0
08551	10	5.19	6.09	3	7	0
08559	7	4.20	6.36	4	2	1
10901	6	1.12	1.50	6	0	0
10930	4	2.38	2.47	4	0	0
10950	11	1.73	2.13	10	1	0
10952	4	1.22	1.26	4	0	0
10956	10	2.39	3.53	6	4	0
10970	5	1.74	1.96	5	0	0
10977	6	1.53	1.66	6	0	0
10987	8	3.50	4.63	6	2	0
10990	9	3.46	4.75	5	4	0

08822	54	5.37	8.45	22	27	5	12180	5	4.10	4.89	3	2	0
08825	15	5.23	14.01	9	4	2	12186	4	6.31	8.02	1	3	0
08826	26	8.86	24.17	8	11	7	12309	4	2.26	4.24	3	1	0
08827	22	6.81	20.04	10	5	7	12401	7	3.13	4.88	5	2	0
08829	20	13.75	23.50	1	13	6	12533	5	4.43	5.08	3	2	0
08833	75	5.71	9.64	28	40	7	12582	5	1.86	2.33	4	1	0
08836	7	3.20	8.48	6	0	1	12590	4	2.40	2.90	3	1	0
08840	4	2.61	3.59	2	2	0	12603	5	3.48	3.69	4	1	0
08848	19	3.72	9.91	13	4	2	13031	4	12.43	14.39	0	3	1
08853	19	2.62	5.43	8	10	1	13066	4	3.02	3.29	2	2	0
08854	7	4.08	7.57	3	3	1	13104	7	5.94	9.29	3	3	1
08858	4	6.54	10.20	2	1	1	13108	6	2.97	11.47	5	0	1
08865	31	5.74	10.58	14	12	5	13215	19	8.01	12.29	6	10	3
08867	17	3.18	4.96	11	5	1	13219	4	6.80	7.56	1	3	0
08873	11	2.97	3.65	7	4	0	13440	4	6.46	11.13	2	1	1
08876	26	2.78	4.69	19	6	1	14075	9	1.66	3.02	6	3	0
08880	5	1.21	1.25	5	0	0	14219	4	1.30	1.49	4	0	0
08889	29	5.48	9.37	10	16	3	14850	6	2.77	5.86	3	2	1
08904	5	7.81	23.90	2	2	1	15122	4	3.45	8.33	2	1	1
10465	7	1.05	1.35	7	0	0	15206	4	2.59	2.77	3	1	0
10509	5	1.06	1.18	5	0	0	15217	6	2.38	3.39	4	2	0
10512	31	3.30	4.73	17	13	1	16335	5	1.74	2.11	4	1	0
10516	4	2.87	4.36	3	1	0	16507	4	2.39	5.47	3	1	0
10520	9	2.71	3.90	7	2	0	16801	5	11.68	13.43	0	4	1
10524	4	2.86	3.58	3	1	0	17003	6	14.93	22.59	1	3	2
10536	9	2.28	6.11	7	1	1	17011	40	11.33	19.51	7	21	12
10538	4	2.61	3.48	3	1	0	17013	13	10.47	15.33	2	6	5
10541	13	2.50	2.89	10	3	0	17033	5	10.03	19.93	0	4	1
10562	9	1.06	1.34	9	0	0	17042	40	9.75	13.63	9	24	7
10566	24	3.16	5.65	16	5	3	17055	45	16.71	24.19	2	23	20
10579	11	3.22	4.40	7	4	0	17067	4	13.70	15.28	0	3	1
10583	11	1.76	2.56	10	1	0	17070	6	7.68	10.19	1	4	1

Zip	Total	Mean	Avg.	–>4	4–>20	20–>
10589	5	4.26	4.83	2	3	0
10598	10	4.31	15.56	6	0	4
10708	4	1.98	2.60	3	1	0
17111	4	13.46	19.71	1	2	1
17112	5	7.72	9.80	1	4	0
17201	5	6.18	9.76	2	2	1
17225	4	12.90	17.89	0	3	1
17402	5	11.80	12.98	0	5	0
17512	5	17.50	24.84	0	2	3
17516	4	11.53	17.83	1	1	2
17517	5	5.12	7.48	1	4	0
17520	8	10.43	13.24	2	4	2
17522	15	5.40	10.14	8	3	4
17532	5	9.87	34.11	1	3	1
17538	8	12.24	15.43	0	5	3
17540	4	15.88	28.99	0	3	1
17543	24	19.54	29.97	2	12	10
17545	4	13.77	21.51	0	2	2
17557	8	12.26	19.72	1	4	3
17584	5	30.14	44.99	0	2	3
17601	67	9.63	15.99	14	36	17
17602	5	6.89	8.36	1	4	0
17603	25	9.24	15.21	2	18	5
18014	6	14.59	29.28	1	2	3
18015	19	7.87	12.69	4	13	2
18017	104	9.80	15.37	16	70	18
18018	30	5.44	9.39	11	14	5
18031	8	30.42	83.44	1	3	4
18034	10	10.14	25.96	3	4	3
18036	22	10.69	21.06	3	13	6
17073	6	17.90	69.56	1	4	1
17078	5	27.61	41.20	0	1	4
17110	6	9.06	11.53	1	5	0
18105	4	5.71	7.74	1	3	0
18106	16	10.47	17.25	3	11	2
18301	4	5.34	7.50	1	3	0
18328	11	1.88	2.31	10	1	0
18901	10	5.68	9.33	4	3	3
18914	4	8.19	8.25	0	4	0
18951	7	8.62	22.46	2	3	2
19010	4	6.10	9.45	1	2	1
19035	4	3.16	3.71	2	2	0
19047	4	0.90	1.17	4	0	0
19067	10	5.78	10.62	3	5	2
19085	4	2.34	2.65	3	1	0
19087	9	5.76	13.14	6	2	1
19317	4	1.32	1.41	4	0	0
19335	8	11.12	24.36	2	4	2
19341	6	3.46	5.32	4	2	0
19355	4	5.21	6.12	1	3	0
19403	6	5.46	8.26	3	2	1
19425	4	8.87	11.39	1	2	1
19426	5	3.35	5.19	3	2	0
19438	6	2.17	3.41	5	1	0
19446	4	3.02	3.52	2	2	0
19464	39	3.51	4.81	25	14	0
19508	8	3.48	11.07	5	2	1
19512	118	3.36	7.97	71	35	12
19518	18	2.79	3.91	13	5	0
19525	23	3.00	7.02	15	7	1

18037	4	9.44	9.72	0	4	0	19533	5	32.27	36.41	0	2	3
18041	5	3.63	5.03	2	3	0	19547	6	15.80	28.26	2	1	3
18042	102	8.42	49.15	33	49	20	19560	5	6.88	7.47	1	4	0
18049	29	7.10	14.31	9	12	8	19604	4	12.97	18.40	0	3	1
18052	10	6.23	7.34	2	8	0	19606	30	7.29	11.03	6	19	5
18054	4	2.49	3.41	3	1	0	19607	6	4.03	7.85	3	2	1
18055	20	13.63	33.83	3	11	6	19610	5	5.61	7.01	2	3	0
18062	28	9.29	13.63	3	22	3	20007	4	1.92	2.45	3	1	0
18064	11	11.92	20.45	1	7	3	20009	6	1.23	1.46	6	0	0
18066	4	18.40	25.82	0	2	2	20015	5	0.90	1.20	5	0	0
18067	5	6.50	11.76	2	2	1	20016	14	2.28	2.80	10	4	0
18069	14	23.12	40.54	2	4	8	20036	6	0.45	0.68	6	0	0
18077	6	5.14	7.16	2	4	0	20705	5	1.39	1.52	5	0	0
18078	15	22.36	45.83	3	3	9	20706	7	0.73	0.93	7	0	0
18102	21	7.52	12.18	6	14	1	20707	11	2.33	4.60	9	2	0
18103	113	6.50	10.02	35	63	15	20708	4	0.87	0.96	4	0	0
18104	92	6.27	10.17	30	52	10	20740	7	1.87	2.43	5	2	0
20748	5	0.77	0.86	5	0	0	21754	7	10.22	12.24	1	5	1
20759	4	4.69	5.17	1	3	0	21755	4	2.26	2.50	3	1	0
20770	8	2.80	5.31	4	4	0	21769	8	2.04	2.52	7	1	0
20777	4	4.50	6.96	3	1	0	21770	15	9.75	12.80	2	10	3
20783	4	1.16	1.38	4	0	0	21771	74	11.43	19.69	15	40	19
20814	11	2.05	2.26	10	1	0	21784	7	6.38	7.01	1	6	0
20815	24	2.30	2.95	20	4	0	21797	5	7.54	11.26	1	3	1
20816	11	4.16	5.26	4	7	0	22003	24	2.42	3.05	18	6	0
20817	29	1.91	3.90	21	7	1	22015	26	2.60	3.58	19	7	0
20832	12	2.48	5.33	10	1	1	22020	17	2.50	3.25	12	5	0
20850	18	3.57	4.04	11	7	0	22021	6	1.39	1.54	6	0	0
20852	19	2.56	3.57	15	4	0	22030	13	2.66	3.13	9	4	0
20853	17	2.98	4.83	11	5	1	22031	8	3.34	4.96	4	4	0
20854	62	3.08	5.06	40	20	2	22032	13	3.96	4.69	5	8	0
20855	18	4.56	6.08	9	8	1	22033	6	3.80	9.45	4	1	1

Zip	Total	Mean	Avg.	–>4	4–>20	20–>
20871	10	9.80	17.59	3	4	3
20872	37	8.19	17.24	11	18	8
20874	16	4.05	5.34	6	10	0
20877	12	4.27	5.55	5	7	0
20878	35	5.20	21.30	18	14	3
20879	47	5.12	10.66	22	20	5
20895	10	1.96	2.77	8	2	0
20901	18	3.94	4.66	11	7	0
20902	20	3.16	4.17	12	8	0
20903	7	2.99	4.04	5	2	0
20904	27	4.06	4.79	13	14	0
20906	30	2.29	3.20	23	7	0
20910	15	2.94	3.51	11	4	0
20912	11	2.48	2.87	8	3	0
21029	6	10.52	11.79	0	6	0
21043	27	8.86	12.21	4	18	5
21044	15	6.33	8.09	2	12	1
21045	18	8.23	9.79	3	15	0
21046	8	9.14	12.07	1	5	2
21048	4	9.14	13.05	1	2	1
21093	8	4.59	5.81	4	4	0
21146	5	1.34	1.71	5	0	0
21157	15	17.99	43.70	1	8	6
21204	4	5.13	7.03	2	2	0
21208	5	2.78	4.68	3	2	0
21212	4	6.20	9.33	1	2	1
21228	5	1.65	2.18	4	1	0
21701	36	5.35	10.75	13	18	5
21740	4	8.31	9.61	1	3	0
22310	7	1.33	1.50	7	0	0

Zip	Total	Mean	Avg.	–>4	4–>20	20–>
22039	7	3.47	5.49	4	3	0
22042	6	0.93	1.07	6	0	0
22043	5	2.49	5.51	4	0	1
22044	5	3.90	4.34	2	3	0
22046	5	1.78	1.86	5	0	0
22066	9	3.74	4.93	6	3	0
22070	12	4.55	19.97	6	4	2
22071	28	3.46	14.19	19	7	2
22075	8	5.26	6.80	3	5	0
22091	28	3.75	5.61	14	13	1
22101	17	2.59	3.74	12	5	0
22102	4	3.58	4.23	3	1	0
2210	18	2.35	2.96	14	4	0
22111	8	5.76	7.69	4	4	0
22115	4	2.48	3.17	3	1	0
22124	6	4.28	7.91	4	1	1
22151	5	2.15	2.28	5	0	0
22152	14	2.29	3.24	9	5	0
22153	16	2.70	3.95	11	5	0
22170	5	4.06	5.57	3	2	0
22180	24	4.11	5.95	12	10	2
22192	8	2.64	4.83	6	1	1
22201	4	1.18	1.43	4	0	0
22203	4	2.60	3.23	3	1	0
22204	4	1.82	1.92	4	0	0
22205	6	1.54	1.60	6	0	0
22206	4	2.01	2.04	4	0	0
22207	14	2.07	3.12	10	4	0
22304	5	1.19	1.31	5	0	0
60061	4	2.64	3.08	3	1	0

22312	7	2.83	3.04	5	2	0	60062	7	2.07	2.37	7	0	0
22601	5	7.69	8.30	0	5	0	60067	20	2.56	3.40	14	6	0
25425	5	12.28	15.15	0	4	1	60068	11	2.16	3.37	9	2	0
28734	6	2.14	2.30	5	1	0	60076	4	2.23	2.36	4	0	0
32201	4	0.79	0.88	4	0	0	60077	4	2.55	3.28	3	1	0
37027	12	2.98	5.29	6	5	1	60089	4	2.23	2.82	3	1	0
37064	10	3.44	8.28	6	3	1	60091	6	2.13	3.73	4	2	0
37122	5	2.76	3.30	3	2	0	60093	8	1.03	1.54	7	1	0
37203	5	9.72	13.80	1	2	2	60101	4	3.82	5.61	2	2	0
37205	7	6.72	8.57	1	5	1	60108	8	2.68	3.49	5	3	0
37215	9	7.68	15.16	3	5	1	60126	21	1.71	2.40	18	3	0
44121	4	0.95	1.37	4	0	0	60134	4	5.37	7.02	2	2	0
44136	17	2.11	2.38	16	1	0	60137	12	3.77	5.44	8	4	0
45385	5	3.70	4.42	3	2	0	60148	4	2.34	2.68	3	1	0
45419	6	7.25	8.03	1	5	0	60172	9	3.62	5.08	5	4	0
45459	5	1.76	2.15	4	1	0	60174	4	1.99	2.00	4	0	0
46220	5	8.90	14.79	1	2	2	60181	5	3.21	4.23	4	1	0
46923	7	22.26	27.57	0	4	3	60187	10	3.67	4.50	5	5	0
47126	10	4.23	7.79	6	3	1	60190	4	6.14	7.28	2	2	0
47130	11	5.13	6.17	5	6	0	60193	5	2.06	2.26	5	0	0
47143	7	5.75	13.75	4	1	2	60194	4	1.05	1.08	4	0	0
47150	19	2.34	2.86	13	6	0	60302	5	2.65	4.26	3	2	0
47172	7	6.40	7.99	1	6	0	60304	4	4.12	4.19	2	2	0
47401	4	3.07	3.29	3	1	0	60305	4	5.46	6.27	1	3	0
48103	5	2.01	2.22	5	0	0	60422	4	2.02	2.18	4	0	0
48104	6	2.64	3.38	4	2	0	60441	5	2.93	3.05	4	1	0
48105	5	1.84	1.98	5	0	0	60463	5	9.21	11.47	1	4	0
49504	4	2.01	2.04	4	0	0	60510	4	7.73	14.83	2	0	2
49506	6	1.05	1.32	6	0	0	60515	10	1.66	3.21	7	3	0
49508	7	0.92	0.99	7	0	0	60521	72	2.87	4.43	48	22	2
55811	5	1.92	2.25	4	1	0	60525	11	3.20	4.24	6	5	0
60004	33	1.99	2.87	24	9	0	60540	13	4.43	5.39	7	6	0

Zip	Total	Mean	Avg.	–>4	4–>20	20–>	Zip	Total	Mean	Avg.	–>4	4–>20	20–>
60005	13	2.12	2.66	10	3	0	60558	9	2.60	3.01	6	3	0
60010	9	4.90	7.21	4	5	0	60559	5	2.73	2.98	4	1	0
60016	4	6.06	17.30	2	1	1	60565	5	3.89	4.81	4	1	0
60025	7	2.79	3.45	6	1	0	60625	10	2.66	3.06	8	2	0
60035	6	2.30	2.66	5	1	0	60645	4	0.72	0.85	4	0	0
60044	4	3.92	6.81	3	0	1	60646	7	1.32	1.42	7	0	0
60045	5	2.13	2.29	5	0	0	60648	4	1.08	1.30	4	0	0
60047	6	2.48	3.02	4	2	0	63628	5	2.10	2.63	4	1	0
60048	4	3.80	4.17	3	1	0	66215	6	3.14	4.05	3	3	0
60053	23	1.35	2.29	19	4	0	72601	12	1.62	2.24	10	2	0
60056	13	1.76	2.39	11	2	0	80302	8	5.05	6.71	4	3	1
80303	7	4.56	6.99	3	3	1	82520	4	7.01	7.56	0	4	0
80501	5	4.03	4.25	2	3	0	82601	4	5.26	5.70	1	3	0
80526	4	10.20	12.76	0	3	1	82801	9	7.51	7.94	0	9	0
80906	13	11.67	16.23	1	8	4	83705	5	3.43	3.75	4	1	0
80918	8	1.98	2.11	8	0	0	83706	6	2.39	3.18	5	1	0
80919	5	17.25	20.43	0	1	4	83835	4	3.48	4.83	2	2	0
82001	5	2.90	3.01	5	0	0	99208	12	6.83	11.16	6	3	3

TABLE 3–D. ZIP CODES BY 3 DIGITS FOR NON-BASEMENTS (ONLY GROUPS WITH COUNTS > = 4 ARE PRINTED.)

Zip	Total	Mean	Avg.	–>4	4–>20	20–>	Zip	Total	Mean	Avg.	–>4	4–>20	20–>
00000	276	1.03	2.82	241	27	8	04800	19	1.05	1.69	18	1	0
01000	51	0.83	1.09	51	0	0	04900	30	1.21	1.70	28	2	0
01100	17	0.67	0.88	16	1	0	05000	11	1.26	1.69	10	1	0
01200	47	0.80	0.97	47	0	0	05100	9	0.96	1.24	9	0	0
01300	11	1.23	1.56	10	1	0	05200	4	1.73	1.96	4	0	0
01400	37	1.69	3.29	28	8	1	05300	8	1.02	1.29	8	0	0
01500	21	1.53	2.30	17	4	0	05400	16	0.74	0.87	16	0	0
01700	87	1.41	2.14	78	8	1	05700	16	0.87	1.21	16	0	0
01800	86	1.20	1.97	78	7	1	06000	86	1.09	1.64	81	5	0
01900	61	1.38	1.98	52	9	0	06100	29	0.70	0.88	29	0	0
02000	56	0.94	1.30	54	2	0	06200	22	1.95	2.79	16	6	0
02100	130	0.90	1.18	126	4	0	06300	21	1.12	1.34	21	0	0
02300	16	1.26	1.76	15	1	0	06400	116	1.33	2.18	101	15	0
02500	4	0.72	0.85	4	0	0	06500	22	1.27	1.93	20	2	0
02600	11	0.78	0.94	11	0	0	06600	8	1.25	2.32	7	1	0
02700	19	1.31	4.47	17	1	1	06700	76	1.24	1.97	69	7	0
02800	62	1.33	2.05	54	8	0	06800	156	1.17	2.07	139	16	1
02900	12	0.88	1.05	12	0	0	06900	20	1.42	2.47	18	2	0
03000	82	1.69	2.38	72	10	0	07000	1274	0.92	1.44	1217	52	5
03100	19	2.18	5.39	14	4	1	07100	32	1.20	2.27	29	2	1
03200	45	1.17	1.84	40	5	0	07200	16	0.59	0.66	16	0	0
03300	5	1.51	1.72	5	0	0	07300	14	1.00	1.75	13	1	0
03400	51	1.57	3.18	41	9	1	07400	2472	1.24	2.00	2226	232	14
03500	14	1.19	2.83	13	0	1	07500	47	0.79	1.42	43	4	0
03600	10	1.51	1.92	9	1	0	07600	319	0.80	1.18	302	17	0

Zip	Total	Mean	Avg.	–>4	4–>20	20–>
03700	19	1.44	2.50	16	3	0
03800	58	2.42	3.90	42	15	1
03900	4	1.42	1.50	4	0	0
04000	76	2.10	3.52	59	16	1
04100	9	1.18	1.45	9	0	0
04200	40	1.53	2.45	33	7	0
04300	15	1.50	3.89	13	1	1
04400	29	0.87	1.26	27	2	0
04500	27	1.20	1.91	24	3	0
04600	37	1.02	1.45	34	3	0
04700	9	1.13	1.32	9	0	0
10000	23	0.77	1.28	21	2	0
10300	50	0.61	0.71	50	0	0
10400	27	0.61	0.69	27	0	0
10500	731	1.36	2.28	641	83	7
10600	10	1.18	1.37	10	0	0
10700	30	0.86	1.03	30	0	0
10800	7	1.02	1.34	7	0	0
10900	3000	1.08	1.57	283	17	0
11000	15	0.83	1.03	15	0	0
11200	31	0.62	0.69	31	0	0
11300	27	0.50	0.54	27	0	0
11400	13	0.60	0.74	13	0	0
11500	36	0.59	0.76	35	1	0
11700	61	0.61	0.67	61	0	0
11800	7	0.61	0.72	7	0	0
11900	8	1.12	1.61	7	1	0
12000	46	1.30	1.66	43	3	0
12100	34	1.58	2.37	27	7	0
12200	9	0.74	0.94	9	0	0
07700	79	1.05	1.59	74	5	0
07800	3624	1.61	2.96	3050	513	61
07900	2108	1.49	3.28	1750	309	49
08000	55	1.16	1.50	51	4	0
08100	4	1.18	1.28	4	0	0
08200	5	0.43	0.43	5	0	0
08500	261	1.62	2.96	221	36	4
08600	71	1.68	4.97	55	12	4
08700	17	0.72	0.90	17	0	0
08800	1759	2.32	8.30	1276	387	96
08900	35	1.39	2.63	30	4	1
15100	79	1.95	3.11	64	14	1
15200	246	1.68	2.96	194	51	1
15300	34	1.61	3.18	29	3	2
15400	10	2.04	3.37	8	2	0
15500	10	3.38	12.20	5	3	2
15600	57	1.99	3.24	41	16	0
15700	17	0.87	1.41	15	2	0
15800	13	2.05	2.64	10	3	0
15900	11	1.96	4.11	9	1	1
16000	26	3.41	7.23	17	6	3
16100	24	1.51	2.29	23	1	0
16200	15	2.61	3.69	10	5	0
16300	26	1.02	1.65	24	2	0
16400	10	1.75	2.34	8	2	0
16500	9	1.20	4.69	7	1	1
16600	15	1.41	2.11	12	3	0
16800	34	5.47	8.21	13	17	4
16900	11	1.56	2.16	10	1	0
17000	359	4.71	8.16	154	172	33

12300	13	0.70	0.76	13	0	0	17100	45	3.93	7.21	22	19	4
12400	50	1.29	1.83	45	5	0	17200	72	3.29	5.59	42	27	3
12500	170	1.60	2.45	136	34	0	17300	55	3.61	7.59	31	15	9
12600	23	1.39	2.78	20	2	1	17400	39	3.93	11.72	22	10	7
12700	18	0.88	1.17	17	1	0	17500	284	4.88	9.93	126	117	41
12800	21	0.72	0.95	20	1	0	17600	163	4.02	7.05	79	76	8
12900	8	0.56	0.65	8	0	0	17700	25	3.94	20.70	15	5	5
13000	44	1.73	3.64	35	8	1	17800	43	1.66	2.86	34	9	0
13100	39	1.89	3.25	30	9	0	17900	20	3.60	8.60	14	5	1
13200	44	2.29	6.04	30	13	1	18000	1093	3.72	10.72	595	392	106
13300	10	1.02	1.46	10	0	0	18100	509	3.42	6.24	276	210	23
13400	21	2.72	4.26	14	7	0	18200	31	2.10	5.17	24	4	3
13600	22	0.83	1.05	21	1	0	18300	78	1.58	4.05	64	10	4
13700	28	1.81	2.57	20	8	0	18400	66	1.34	2.21	59	6	1
13800	30	1.59	2.15	27	3	0	18500	5	1.54	2.47	4	1	0
13900	16	0.91	1.60	14	2	0	18600	20	1.58	1.89	19	1	0
14000	56	1.13	2.75	50	4	2	18700	12	1.65	2.61	10	2	0
14100	7	1.62	4.80	6	0	1	18800	15	1.17	1.66	15	0	0
14200	20	0.74	1.18	18	2	0	18900	147	2.38	4.19	110	29	8
14400	21	1.47	2.37	19	2	0	19000	250	1.72	3.36	209	35	6
14500	11	1.00	1.14	11	0	0	19100	50	1.03	1.41	46	4	0
14600	19	1.12	1.37	19	0	0	19300	138	1.80	4.00	115	20	3
14700	10	2.46	5.48	6	3	1	19400	196	2.06	4.96	164	23	9
14800	30	1.45	2.07	25	5	0	19500	191	3.68	8.33	111	57	23
15000	92	2.55	6.23	62	25	5	19600	96	4.19	7.66	42	46	8
19700	4	0.86	0.95	4	0	0	27200	16	1.25	1.60	15	1	0
19800	12	0.96	1.38	11	1	0	27300	10	1.32	1.61	9	1	0
19900	9	0.60	0.67	9	0	0	27400	11	1.30	1.69	10	1	0
20000	163	0.90	1.23	155	8	0	27500	16	1.58	1.91	16	0	0
20600	59	1.28	1.72	52	7	0	27600	29	1.44	1.94	26	3	0
20700	297	1.51	2.60	254	38	5	27700	7	1.45	1.78	7	0	0
20800	1122	1.92	2.96	909	204	9	27800	9	0.67	0.86	9	0	0

Zip	Total	Mean	Avg.	->4	4->20	20->
20900	311	1.62	2.55	270	40	1
21000	295	2.75	4.44	185	103	7
21100	77	2.75	6.68	47	26	4
21200	73	1.23	2.02	65	7	1
21400	23	1.56	2.13	22	1	0
21500	11	2.28	4.16	7	3	1
21600	18	1.05	2.06	16	2	0
21700	292	3.84	8.00	149	124	19
21900	5	0.95	1.10	5	0	0
22000	658	1.71	2.76	554	99	5
22100	423	1.68	2.46	358	62	3
22200	82	1.20	1.63	78	4	0
22300	134	1.08	1.42	131	3	0
22400	13	1.15	1.21	13	0	0
22500	6	1.60	2.03	5	1	0
22600	22	2.42	4.07	15	6	1
22700	7	1.01	1.07	7	0	0
22800	7	1.70	2.22	6	1	0
22900	16	1.68	3.85	13	2	1
23000	9	1.20	1.43	9	0	0
23100	11	1.89	2.40	9	2	0
23200	29	2.25	4.01	21	7	1
23400	10	0.70	0.77	10	0	0
23600	12	0.74	0.81	12	0	0
24000	27	2.78	3.81	18	9	0
24100	12	1.12	1.30	12	0	0
24300	4	1.39	1.53	4	0	0
24400	12	2.82	7.52	10	1	1
24500	32	2.13	2.96	27	5	0
25400	40	2.22	5.71	29	6	5

Zip	Total	Mean	Avg.	->4	4->20	20->
28000	22	2.39	3.48	17	5	0
28100	9	0.99	1.18	9	0	0
28200	11	1.05	1.29	11	0	0
28300	6	1.33	1.77	6	0	0
28400	4	0.72	0.75	4	0	0
28600	18	2.26	3.53	14	4	0
28700	95	2.92	4.57	62	31	2
28800	8	2.75	4.12	5	3	0
28900	6	3.55	6.84	4	1	1
29100	4	1.66	2.22	4	0	0
29200	4	0.75	0.88	4	0	0
29400	12	0.63	0.73	12	0	0
29500	11	1.19	1.29	11	0	0
29600	14	1.94	2.87	10	4	0
30000	35	1.74	2.07	33	2	0
30100	10	1.61	3.05	8	2	0
30200	13	1.79	2.36	11	2	0
30300	35	2.10	2.63	30	5	0
30500	11	2.61	3.30	9	2	0
30600	5	2.94	3.36	3	2	0
30700	5	1.03	1.38	5	0	0
30900	5	1.05	1.22	5	0	0
31400	11	0.77	1.01	11	0	0
31500	5	0.67	0.83	5	0	0
32000	15	0.79	0.97	15	0	0
32200	50	0.72	0.84	49	1	0
32300	6	1.60	2.87	5	1	0
32500	15	1.02	1.16	15	0	0
32600	31	2.06	3.48	25	6	0
32700	17	1.27	2.04	14	3	0

26000	10	2.64	5.71	6	3	1
26100	5	1.90	1.96	5	0	0
26300	12	1.98	2.63	8	4	0
26500	10	1.31	1.57	10	0	0
26700	5	1.46	2.63	4	1	0
27000	4	1.92	2.53	3	1	0
27100	10	5.45	8.77	4	5	1
33700	5	1.33	7.25	4	0	1
33800	82	2.50	5.18	56	24	2
33900	4	3.12	4.00	3	1	0
34200	13	1.45	2.85	9	4	0
35000	8	1.26	1.47	8	0	0
35100	9	1.67	1.93	9	0	0
35200	12	1.92	2.68	11	1	0
35600	4	2.92	4.65	2	2	0
35800	18	4.16	7.01	9	8	1
37000	388	2.43	3.47	297	92	5
37100	42	1.71	2.05	40	2	0
37200	196	2.47	3.73	135	58	3
37300	50	1.69	3.28	45	4	1
27400	17	1.45	1.92	15	2	0
37600	29	3.57	4.82	15	14	0
37700	12	2.12	2.87	9	3	0
37800	46	3.21	5.39	29	15	2
37900	42	2.00	3.02	32	10	0
38100	5	1.19	1.29	5	0	0
38400	6	2.80	3.27	5	1	0
38500	17	2.07	6.87	12	4	1
39400	4	1.73	2.04	4	0	0
40000	6	3.06	6.00	4	1	1
40100	6	1.58	1.75	6	0	0
40200	30	2.83	4.04	20	10	0
32900	6	0.58	0.61	6	0	0
33000	7	1.03	1.40	7	0	0
33100	6	1.85	2.33	5	1	0
33300	12	0.84	1.01	12	0	0
33400	21	0.82	1.28	20	1	0
33500	128	1.31	1.97	119	8	1
33600	41	0.94	1.12	41	0	0
44500	4	1.40	1.47	4	0	0
44600	15	2.48	5.32	10	5	0
44700	7	3.71	4.73	4	3	0
44800	16	1.56	2.51	13	3	0
44900	4	3.18	4.06	3	1	0
45000	8	2.69	3.84	6	2	0
45100	5	3.73	4.08	3	2	0
45200	22	1.66	2.80	18	4	0
45300	44	3.06	5.06	27	16	1
45400	63	2.79	3.99	43	19	1
45500	4	1.85	2.18	3	1	0
45600	11	2.03	3.05	8	3	0
45700	9	1.45	1.63	9	0	0
45800	7	1.97	3.48	5	2	0
46000	13	3.81	6.39	8	4	1
46100	10	1.24	1.51	10	0	0
46200	35	4.15	6.81	18	13	4
46300	41	1.80	2.93	36	4	1
46400	7	0.84	1.16	7	0	0
46500	6	1.17	1.21	6	0	0
46800	5	3.43	10.66	2	2	1
46900	12	3.74	11.08	6	4	2
47100	89	2.46	6.04	65	19	5
47300	7	3.25	3.87	4	3	0
47400	8	1.72	2.23	6	2	0

Zip	Total	Mean	Avg.	–>4	4–>20	20–>
40300	10	4.29	5.24	2	8	0
40400	14	2.05	3.84	9	5	0
40500	27	4.45	6.80	11	15	1
41000	12	1.83	3.27	11	1	0
42000	4	1.04	1.46	4	0	0
42100	6	1.38	3.29	4	2	0
43000	32	2.99	3.82	20	12	0
43100	14	3.80	5.05	6	8	0
43200	52	4.47	6.61	24	25	3
43400	4	1.22	1.41	4	0	0
43500	9	2.24	3.08	7	2	0
43600	8	1.32	1.54	8	0	0
43700	6	1.45	1.88	6	0	0
43900	5	4.42	5.50	2	3	0
44000	42	1.16	1.61	38	4	0
44100	65	1.21	2.59	62	2	1
44200	35	1.78	2.31	30	5	0
44300	15	2.19	2.73	12	3	0
44400	15	2.59	5.06	10	4	1
49900	9	0.91	1.76	8	1	0
50000	8	2.92	4.16	4	4	0
50100	4	2.25	3.29	3	1	0
50300	4	3.88	6.05	2	2	0
50600	6	3.02	3.96	3	3	0
51000	4	6.22	7.82	1	3	0
51300	7	3.13	6.11	4	2	1
52000	6	1.67	2.26	5	1	0
52100	5	2.28	3.01	4	1	0
52200	4	2.45	2.74	3	1	0
52300	4	2.82	3.94	2	2	0
47800	29	2.67	3.34	22	7	0
47900	7	2.78	3.52	5	2	0
48000	76	1.19	1.57	73	3	0
48100	92	1.62	2.14	83	9	0
48200	17	0.83	0.94	17	0	0
48400	20	1.27	2.06	17	3	0
48500	5	0.60	0.75	5	0	0
48600	8	1.06	1.15	8	0	0
48700	6	0.50	0.58	6	0	0
48800	34	1.30	1.66	32	2	0
48900	8	2.25	2.37	7	1	0
49000	58	1.45	2.12	49	9	0
49100	5	0.94	1.17	5	0	0
49200	7	2.18	3.81	5	2	0
49300	27	1.13	1.81	25	2	0
49400	41	1.37	2.04	39	2	0
49500	75	1.03	1.25	74	1	0
49700	5	1.65	3.10	4	1	0
49800	12	1.46	1.66	11	1	0
59000	4	294	3.67	2	2	0
59600	4	3.02	5.17	2	2	0
59700	11	2.93	3.23	6	5	0
59800	13	2.71	4.24	10	3	0
59900	4	1.93	2.12	4	0	0
60000	513	1.39	1.88	467	45	1
60100	265	1.68	2.38	225	39	1
60200	17	0.65	0.84	16	1	0
60300	11	1.23	1.71	10	1	0
60400	166	1.47	2.64	142	23	1
60500	251	2.17	3.15	187	64	0

52700	7	2.45	2.77	5	2	0
52800	4	3.08	4.33	3	1	0
53000	25	1.90	2.64	20	5	0
53100	17	1.66	3.03	13	3	1
53200	18	1.14	1.49	16	2	0
53400	4	3.13	4.28	2	2	0
53500	6	1.17	1.59	5	1	0
53700	7	1.65	2.02	6	1	0
53900	5	1.07	1.20	5	0	0
54000	5	1.54	1.75	5	0	0
54100	12	1.07	1.40	11	1	0
54300	6	1.02	1.41	6	0	0
54400	21	4.96	9.60	10	6	5
54500	7	1.43	1.81	6	1	0
54800	12	1.60	1.75	12	0	0
54900	23	2.10	3.20	19	4	0
55000	14	2.32	3.11	11	3	0
55100	18	1.64	2.90	15	3	0
55300	28	2.09	2.75	22	6	0
55400	31	1.73	2.43	26	5	0
55700	11	2.22	2.58	10	1	0
55800	11	1.40	1.93	10	1	0
56000	13	4.19	5.13	7	6	0
56300	18	1.46	2.08	15	3	0
56400	4	2.12	2.29	4	0	0
56500	14	1.87	3.32	11	3	0
56600	5	3.29	4.43	3	2	0
57700	7	1.87	2.61	4	3	0
58100	15	2.24	4.29	9	6	0
58200	8	3.83	5.13	3	5	0
58300	6	2.26	3.06	5	1	0
58400	6	1.86	2.46	5	1	0
60600	285	1.49	2.35	242	41	2
60900	10	1.48	1.78	10	0	0
61000	8	3.44	4.46	5	3	0
61100	4	1.61	1.98	3	1	0
61200	10	1.74	2.16	9	1	0
61300	11	2.15	3.70	10	1	0
61500	7	4.09	5.58	4	3	0
61600	7	3.14	4.52	4	3	0
61700	9	4.82	8.64	4	4	1
61800	12	1.49	2.06	11	1	0
62000	8	1.48	1.89	7	1	0
62200	15	1.69	2.00	13	2	0
62500	5	3.92	4.76	3	2	0
62600	15	3.23	5.15	8	6	1
62700	5	2.91	4.40	3	2	0
62800	6	1.84	1.97	6	0	0
62900	5	0.59	0.72	5	0	0
63000	42	1.25	1.63	41	1	0
63100	59	1.53	2.10	51	8	0
63300	13	1.69	1.95	12	1	0
63600	14	1.26	1.85	11	3	0
63700	4	2.77	3.62	2	2	0
64000	26	1.89	4.15	21	4	1
64100	40	2.03	4.25	30	8	2
64400	5	0.98	1.31	5	0	0
64800	18	0.86	1.06	18	0	0
65200	7	1.42	1.77	7	0	0
66000	20	2.10	3.19	14	6	0
66200	50	2.89	4.63	33	16	1
66700	7	1.13	1.36	7	0	0
68100	4	4.66	5.63	2	2	0
69000	4	2.74	3.28	3	1	0

Zip	Total	Mean	Avg.	–>4	4–>20	20–>
58700	6	2.09	2.56	5	1	0
70000	5	0.57	0.67	5	0	0
70500	5	1.15	1.59	4	1	0
71900	4	1.39	1.60	4	0	0
72200	6	1.28	1.40	6	0	0
72600	13	1.67	2.39	9	4	0
72700	15	1.38	2.48	14	1	0
73000	5	2.10	2.18	5	0	0
73100	12	1.22	1.53	12	0	0
74000	16	1.21	1.92	14	2	0
74100	7	1.24	1.33	7	0	0
75000	6	1.73	2.59	5	1	0
75200	17	3.14	4.30	12	5	0
76000	5	6.20	114.07	3	1	1
76100	10	3.87	5.06	4	6	0
77000	8	0.85	1.14	8	0	0
78200	7	0.65	0.73	7	0	0
78600	10	3.25	7.78	8	1	1
78700	13	1.59	1.95	11	2	0
80000	17	3.32	4.52	10	7	0
80100	38	3.04	3.59	24	14	0
80200	41	4.46	7.00	17	21	3
80300	57	3.73	5.65	32	22	3
80400	74	3.23	7.01	47	24	3
80500	39	2.98	3.85	26	13	0
80600	4	0.81	0.95	4	0	0
80800	26	4.73	5.98	11	14	1
80900	130	3.25	5.26	83	41	6
81000	5	2.83	3.19	4	1	0
81200	10	3.65	4.40	5	5	0
69300	4	2.57	2.82	4	0	0
83500	97	1.78	2.66	81	15	1
53600	227	1.61	2.38	202	24	1
83700	139	2.13	2.68	113	26	0
83800	187	2.54	5.72	131	50	6
84000	11	2.84	4.23	7	4	0
84100	13	1.84	2.22	12	1	0
84700	4	0.74	0.74	4	0	0
85000	18	1.46	1.72	18	0	0
85200	16	1.81	2.22	15	1	0
85300	12	2.35	4.37	10	1	1
85600	8	1.74	3.00	6	2	0
85700	18	1.86	2.41	15	3	0
86000	10	1.59	2.55	8	2	0
86300	17	2.06	2.67	13	4	0
87000	8	2.71	3.02	7	1	0
87100	20	3.49	16.08	13	4	3
87500	11	2.89	3.73	6	5	0
87800	5	2.20	2.56	4	1	0
89100	4	1.35	1.37	4	0	0
89400	15	3.02	4.29	11	4	0
89500	5	2.18	2.96	4	1	0
90000	18	0.77	0.93	18	0	0
90200	10	0.92	1.17	10	0	0
90600	6	0.88	1.09	6	0	0
91100	7	0.62	078	7	0	0
91300	18	1.79	2.18	16	2	0
91400	7	1.25	1.53	7	0	0
91700	4	0.62	0.64	4	0	0
92000	26	0.93	1.10	26	0	0

81300	4	1.76	2.02	4	0	0
81400	6	3.58	4.21	2	4	0
81500	7	1.51	1.70	7	0	0
81600	6	2.60	3.84	4	2	0
82000	10	1.97	2.31	10	0	0
82300	5	3.32	8.95	4	0	1
82400	7	0.94	1.23	7	0	0
82500	24	5.01	7.50	12	11	1
82600	12	3.95	7.06	5	7	0
82700	4	3.48	4.35	3	1	0
82800	26	2.57	3.38	18	8	0
83000	16	2.74	4.89	12	4	0
83200	44	1.72	2.61	36	8	0
83300	101	2.93	4.69	64	36	1
83400	44	2.26	3.40	36	8	0
95200	4	2.75	3.36	2	2	0
95300	8	1.04	1.28	8	0	0
95400	21	1.01	3.12	20	0	1
95600	6	1.18	3.04	5	1	0
95700	9	1.62	3.12	8	1	0
95900	5	0.73	0.83	5	0	0
96100	8	2.06	2.53	6	2	0
97000	10	1.21	1.57	9	1	0
97200	18	1.65	1.95	17	1	0
97300	10	1.34	2.04	8	2	0
97400	23	1.03	2.46	21	2	0
97500	16	2.45	6.20	11	3	2
98000	16	0.70	1.09	15	1	0
98100	9	0.62	0.71	9	0	0
98200	24	0.47	0.50	24	0	0

92100	5	0.84	0.95	5	0	0
92200	6	0.65	0.77	6	0	0
92300	23	0.86	1.22	22	1	0
92600	16	1.12	1.27	16	0	0
93000	4	1.99	2.15	4	0	0
93400	5	1.78	2.53	4	1	0
93500	6	1.65	1.86	6	0	0
93900	4	0.58	0.67	4	0	0
94000	10	0.82	1.24	9	1	0
94100	6	0.88	1.13	6	0	0
94500	19	0.78	0.93	19	0	0
94700	10	0.89	1.10	10	0	0
94900	12	0.70	0.82	12	0	0
95000	12	1.18	1.40	12	0	0
95100	8	0.64	0.73	8	0	0
98300	12	0.68	0.86	12	0	0
98400	17	1.11	2.73	15	2	0
98500	4	0.60	0.71	4	0	0
98600	5	1.53	1.73	5	0	0
98800	23	2.63	12.68	18	1	4
98900	7	1.17	1.65	7	0	0
99000	16	4.65	9.72	10	1	5
99100	14	2.83	4.24	9	4	1
99200	70	4.24	8.25	35	28	7
99300	10	1.58	1.70	10	0	0
99500	12	0.73	1.05	12	0	0
99600	7	2.43	3.07	4	3	0
99700	5	1.05	1.88	4	1	0
99800	9	0.70	0.81	9	0	0

TABLE 3–E. ZIP CODES BY 3 DIGITS FOR BASEMENTS (ONLY GROUPS WITH COUNTS > = 4 ARE PRINTED.)

Zip	Total	Mean	Avg.	–>4	4–>20	20–>	Zip	Total	Mean	Avg.	–>4	4–>20	20–>
00000	89	1.97	4.19	64	23	2	07600	100	1.61	2.29	90	9	1
01000	14	1.58	1.90	13	1	0	07700	12	2.12	4.41	7	5	0
01100	4	1.83	2.68	3	1	0	07800	1027	3.51	6.94	582	395	50
01200	6	1.54	1.77	6	0	0	07900	707	3.42	7.65	400	257	50
01400	12	5.36	7.43	5	6	1	08000	11	2.31	2.63	10	1	0
01500	4	2.06	2.18	4	0	0	08500	121	4.32	7.55	62	50	9
01700	34	2.97	9.34	21	12	1	08600	25	3.56	11.82	15	6	4
01800	26	3.92	7.20	15	8	3	08800	628	5.20	13.57	287	253	88
01900	14	3.47	4.65	8	6	0	08900	11	4.07	14.17	6	3	2
02000	24	2.29	3.02	21	3	0	10000	4	1.18	1.58	4	0	0
02100	65	2.18	3.00	55	10	0	10300	4	0.96	1.41	4	0	0
02300	8	1.37	1.41	8	0	0	10400	12	1.16	1.53	12	0	0
02800	14	3.09	3.63	7	7	0	10500	187	2.50	4.57	139	38	10
03000	17	3.27	4.32	9	8	0	10600	4	1.16	1.54	4	0	0
03200	4	2.02	2.12	4	0	0	10700	12	1.94	2.27	11	1	0
03400	13	2.66	3.57	10	3	0	10800	5	3.42	3.80	3	2	0
03700	6	4.77	6.34	2	4	0	10900	86	1.91	2.66	73	13	0
03800	8	1.67	3.04	7	1	0	11200	10	2.30	2.78	7	3	0
04000	8	6.54	12.41	4	3	1	11400	5	1.81	2.09	4	1	0
04400	4	2.64	3.00	3	1	0	11500	8	0.66	0.72	8	0	0
04600	5	3.58	5.44	1	4	0	11700	12	1.38	1.74	11	1	0
04900	7	1.06	1.33	7	0	0	12000	13	5.50	19.26	5	6	2
05400	4	1.76	2.63	3	1	0	12100	13	5.49	9.12	5	7	1
06000	37	2.38	8.33	28	8	1	12300	5	1.89	3.58	4	1	0
06100	9	1.76	4.01	6	2	1	12400	13	3.10	4.44	9	4	0

06200	4	1.74	2.94	3	1	0
06300	4	2.73	3.32	3	1	0
06400	25	2.57	3.83	16	8	1
06500	9	3.22	5.26	5	4	0
06700	19	2.77	4.26	15	3	1
06800	37	3.68	7.32	19	14	4
07000	432	1.74	2.84	376	51	5
07100	15	2.07	4.03	11	3	1
07200	5	2.62	3.09	4	1	0
07400	595	2.16	3.57	455	124	16
07500	15	1.78	3.04	13	2	0
14000	11	1.73	3.11	7	4	0
14200	4	1.30	1.49	4	0	0
14600	7	1.68	1.94	7	0	0
14700	7	4.85	7.34	3	3	1
14800	12	3.74	6.61	5	5	2
15000	11	5.70	11.09	4	6	1
15100	16	3.04	4.63	11	4	1
15200	30	2.76	3.38	22	8	0
15500	5	5.85	9.03	2	3	0
15600	7	2.21	2.37	7	0	0
15800	5	0.78	1.25	5	0	0
15900	4	6.03	12.57	2	1	1
16000	5	4.40	9.99	3	1	1
16100	4	3.00	5.25	3	1	0
16300	6	1.47	1.87	5	1	0
16400	10	1.81	2.84	7	3	0
16500	9	2.18	4.33	6	3	0
16800	5	11.68	13.43	0	4	1
17000	201	11.09	20.11	37	105	59
17100	21	9.15	12.24	4	16	1
17200	18	9.60	16.24	4	11	3
12500	28	2.92	4.33	18	9	1
12600	7	2.31	2.99	6	1	0
12700	8	3.10	4.46	5	3	0
13000	23	2.72	5.32	13	9	1
13100	22	2.77	7.63	16	3	3
13200	30	6.02	9.55	12	15	3
13300	5	6.03	13.84	2	2	1
13400	6	7.17	10.51	2	3	1
13600	4	4.89	8.38	1	3	0
13700	5	2.33	3.97	4	1	0
13800	6	5.24	9.19	2	3	1
20900	128	3.08	3.95	83	45	0
21000	110	6.49	9.24	30	70	10
21100	42	5.60	19.53	18	15	9
21200	40	1.96	3.45	31	8	1
21400	4	3.39	3.70	2	2	0
21700	186	7.68	17.29	55	98	33
22000	250	2.96	6.12	163	79	8
22100	140	3.05	4.53	94	42	4
22200	39	1.78	2.35	34	5	0
22300	32	1.56	1.89	30	2	0
22600	7	5.00	6.44	2	5	0
22700	4	2.54	3.32	3	1	0
22900	5	4.19	5.41	3	2	0
24000	10	2.40	3.25	8	2	0
24100	4	8.31	13.98	1	2	1
24500	9	4.56	6.73	3	6	0
25400	10	12.37	14.74	0	8	2
28700	17	2.79	3.96	11	6	0
30000	5	4.23	4.44	2	3	0
30300	10	3.21	3.50	7	3	0
32200	4	0.79	0.88	4	0	0

Zip	Total	Mean	Avg.	->4	4->20	20->
17300	22	5.30	10.59	11	8	3
17400	11	7.83	9.81	2	9	0
17500	125	10.58	20.29	25	62	38
17600	98	9.46	15.51	17	58	23
17700	5	11.11	17.66	1	2	2
18000	488	9.35	25.87	114	263	111
18100	248	6.66	10.65	76	144	28
18200	6	5.77	13.70	3	2	1
18300	31	3.63	6.41	17	11	3
18400	8	2.28	3.08	6	2	0
18600	7	8.40	15.65	1	5	1
18900	42	4.74	9.50	18	18	6
19000	67	3.55	7.13	40	20	7
19100	14	3.16	5.58	8	6	0
19300	47	5.06	22.70	24	17	6
19400	91	3.65	5.70	55	32	4
19500	213	4.01	9.61	119	70	24
19600	56	6.45	10.52	15	32	9
19800	4	3.80	5.35	2	2	0
20000	52	1.41	2.22	43	9	0
20600	8	2.90	3.17	6	2	0
20700	86	2.08	4.28	65	18	3
20800	406	3.74	8.26	236	144	26
45300	23	6.32	10.09	9	10	4
45400	24	3.45	4.52	13	11	0
46200	11	6.90	10.26	3	6	2
46300	11	1.53	2.14	9	2	0
46500	5	2.40	2.84	4	1	0
46900	7	22.26	27.57	0	4	3
47100	56	3.95	6.47	30	23	3

Zip	Total	Mean	Avg.	->4	4->20	20->
35600	4	15.26	16.62	0	2	2
37000	33	3.39	6.54	18	12	3
37100	11	3.47	4.18	4	7	0
37200	40	5.48	12.72	17	16	7
37300	11	2.96	3.63	7	4	0
37400	4	1.57	1.72	4	0	0
37600	6	5.46	7.50	2	3	1
37800	5	2.66	2.84	4	1	0
37900	4	4.13	4.43	1	3	0
40200	11	4.35	7.18	6	5	0
43000	8	7.99	15.38	3	3	2
43100	4	9.80	23.18	2	0	2
43200	11	7.40	9.02	2	9	0
43500	4	6.44	7.20	1	3	0
43600	4	3.37	17.84	3	0	1
44000	17	1.94	2.99	11	6	0
44100	41	1.56	1.99	39	2	0
44200	9	2.59	3.78	7	2	0
44400	4	1.56	1.66	4	0	0
44600	4	2.42	2.68	3	1	0
44800	5	13.09	24.21	1	3	1
45000	4	6.74	7.94	1	3	0
45200	5	2.12	6.49	4	0	1
60500	153	3.14	4.73	95	54	4
60600	52	1.55	1.92	49	3	0
61300	4	3.50	4.77	2	2	0
61800	5	4.00	4.25	2	3	0
62200	4	2.10	2.35	4	0	0
63000	6	2.93	4.37	3	3	0
63100	23	2.37	2.81	17	6	0

47400	7	4.18	4.66	3	4	0	63300	9	2.21	2.69	8	1	0
48000	15	1.85	2.57	13	2	0	63600	7	2.17	2.55	6	1	0
48100	28	1.93	2.33	25	3	0	64000	6	3.38	4.50	4	2	0
48200	4	1.22	1.34	4	0	0	64100	4	2.90	3.90	3	1	0
48800	11	2.40	2.71	10	1	0	66000	4	3.39	3.82	3	1	0
48900	5	1.90	3.20	4	1	0	66200	21	2.99	4.79	11	9	1
49000	17	2.11	2.86	13	4	0	72600	14	1.52	2.15	12	2	0
49300	8	1.91	2.75	6	2	0	80000	6	6.03	6.57	1	5	0
49400	15	1.36	1.74	15	0	0	80100	10	7.18	8.36	2	8	0
49500	26	1.45	2.29	24	1	1	80200	9	5.46	7.12	4	5	0
49800	5	2.21	3.77	4	1	0	80300	18	4.81	6.60	8	8	2
49900	4	1.16	1.34	4	0	0	80400	5	8.86	26.26	2	2	1
53100	5	2.92	4.59	3	2	0	80500	23	6.53	10.59	6	14	3
53400	5	4.06	4.99	3	2	0	80800	4	6.16	15.35	2	1	1
54300	4	3.43	4.41	2	2	0	80900	42	5.24	9.75	20	13	9
54400	9	5.37	8.69	4	3	2	82000	8	2.30	2.48	8	0	0
54700	5	2.49	2.70	4	1	0	82500	4	7.01	7.56	0	4	0
55100	7	2.81	3.57	5	2	0	82600	10	3.14	3.83	7	3	0
55300	6	4.53	6.83	2	4	0	82800	10	7.09	7.56	0	10	0
55400	10	3.08	4.34	6	4	0	83200	5	5.46	5.78	1	4	0
55800	8	2.68	3.21	5	3	0	83300	7	3.31	3.78	3	4	0
58100	4	3.30	4.61	2	2	0	83400	4	2.45	2.80	3	1	0
60000	232	2.26	3.55	177	51	4	83600	17	2.54	3.27	10	7	0
60100	124	2.87	4.07	53	40	1	83700	17	2.98	3.60	12	5	0
60200	6	0.79	0.92	6	0	0	83800	9	3.41	4.71	4	5	0
60300	13	3.79	4.85	6	7	0	97200	5	1.14	1.33	5	0	0
60400	48	2.80	4.43	34	13	1	99200	21	6.44	11.07	10	7	4

TABLE 3–F. DISTRIBUTION OF RADON CONCENTRATIONS IN CANADIAN HOMES

Some data on Canada, derived from Canadian government studies, are listed in the following table.

Location	Radon Concentration Range, pCi/l											Total
	0.0–0.5	0.6–1.0	1.1–1.5	1.6–2.0	2.1–2.5	2.6–3.0	3.1–3.5	3.6–4.0	4.1–4.5	4.6–5.0	>5.0	
Calgary	552	171	72	36	21	14	7	11	3	0	13	900
Charlottetown	415	147	88	44	29	24	15	9	7	5	90	813
Fredericton	173	95	52	37	34	11	13	9	5	7	19	455
Montreal	425	99	34	10	12	6	2	3	0	3	6	600
Quebec	405	91	36	10	11	6	1	5	7	3	9	584
Saint John	557	96	68	27	26	21	6	11	3	9	42	866
Sherbrooke	516	115	81	44	25	32	7	13	8	4	60	905
St. John's	362	98	40	25	14	11	6	7	5	1	16	585
St. Lawrence	171	69	40	32	19	13	10	10	5	9	54	432
Sudbury	311	161	112	58	46	20	13	13	9	8	21	772
Thunder Bay	280	90	105	48	32	15	13	14	4	5	24	627
Toronto	537	129	53	14	9	4	4	0	0	0	1	751
Vancouver	721	73	24	3	2	0	0	0	0	0	0	823

4

Measuring Radon Levels

(Note: This chapter was written by Richard Greenhaus, Project Manager, Consumers Union.)

Radon has been with us since early humans lived in caves, but our recognition of it as a common threat to life and health is a very recent thing. What we don't know about radon not only could fill many books, papers, and articles, but has already done so. There is a crying need for more and better information on such matters as the precise nature and magnitude of health risks associated with radon: the interactions between radon and other lung-cancer-causing agents, such as cigarette smoke; the mechanisms by which radon enters homes; and the connection between geological formations and radon distribution. At this point, there is no clear picture of how many homes—and which ones—present a radon problem. Although a good deal of research work is now in progress, an enormous amount of vital information is missing.

What *is* known is that a clear, unequivocal relationship exists between exposure to high levels of radon and the incidence of lung cancer. Usual estimates of annual fatalities are based on estimated average exposure levels of 1 pCi/l. The distribution of radon in homes is severely skewed, though, as outdoor background level is about 0.1–0.2 pCi/l, and indoor levels of over 1,000 pCi/l have been measured. As a consequence, a relatively small percentage of homes—or between 1 and 2 million in the U.S.—have radon levels of at least ten times the average.

The risk of exposure to these levels is comparable to the risk of working in a uranium mine. There has been a good deal of widely publicized controversy about exactly how great this risk is, but the argument essentially is whether the risk (at these high exposure levels) is extremely serious or merely very serious.

Considering the great gaps in our precise understanding of radon and its effects, particularly at low exposure levels, it is tempting to suggest waiting until more is known. In light of the seriousness of radon as a threat to health, though, such a suggestion would be irresponsible.

From the standpoint of the homeowner, the real question is, What about *my* house? The answer to this can only be found through testing that particular house. Consumers Union, in the fall of 1986, set out to explore the homeowner's options for making this important measurement.

At that time, public awareness of radon as a threat to health was more limited than it is today, and relatively few radon detection kits were widely distributed. CU chose seven of the most popular models for its tests. We didn't anticipate what would happen in the marketplace over the next few months: radon testing and radon detection devices proliferated explosively. By the time we finished our tests, well over 200 models of detection devices, were being offered. (Many of them are the same products, but distributed, evaluated and reported upon by different companies.)

To a great extent, the problems of radon detection and measurement can be addressed generically; detectors of a given type share common advantages and disadvantages. And while the scope of CU's test project turned out to be extremely limited in terms of the present range of available models, our test program did provide information that was not readily available. We think that our test findings, even if they cover only a small percentage of the radon testers sold, will be helpful to the homeowner seeking guidance.

Testing for Radon

Radon concentrations can't be measured directly. Radon itself can't be seen or smelled. Chemical detection tests aren't possible, since radon doesn't combine with anything. Only indirect evaluations of concentrations of radon—made by measuring radioactivity—are possible. Although radon daughters are chemically reactive solids, and could theoretically be measured directly, they just don't exist long enough (and aren't present in high enough concentrations) to make direct chemical or physical measurements feasible. Measurements, then, are based upon radioactivity.

We found three basic types of radon measurement devices in common use:

1. The most accurate testers are, unfortunately, far too expensive for the individual homeowner to consider. These are electronic laboratory instruments, typically priced at several thousand dollars apiece. Their price tags, and the on-site expertise required to operate and maintain them, have largely limited their use to research organizations, where their accuracy and ability to make continuous readings on an hour-by-hour basis have provided the yardstick against which all of the other detectors are measured. Their other users tend to be commercial establishments out to sell radon mitigation work to the householder: laboratory-type instruments permit "grab sample" measurements to be made in an hour or less. Unfortunately, the value of a grab sample test is almost nil: even if the operator is honest, a single one-hour reading tells you comparatively little about long-term average radon levels. And if he's *not* honest, an experienced operator can make one of these gadgets read almost anything he wants it to, through picking locations where local concentrations of radon are likely to be high (over cracks or floor drains, for example), or through simple misadjustment of the device.

The best-known laboratory instrument for measuring radioactivity—the geiger counter—is unsuitable for radon detection. A geiger counter is most sensitive to penetrating beta and gamma radiation, which are present everywhere, from natural sources such as cosmic rays. Since a geiger counter requires the radiation being measured to pass through the glass walls of its detector, and since alpha particles are intercepted easily by glass (or by virtually any other substance, for that matter), very few of the alpha particles that are of concern end up where they can be counted. A geiger-counter reading consists almost entirely of radiation that is, as far as radon detection goes, irrelevant.

The laboratory instrument CU purchased for its tests is characterized as an *integrating scintillation counter/recorder.* It works on a very simple principle: an alpha particle striking a molecule of zinc sulfide will cause a visible flash of light (or scintillation). The instrument's sensing cell consists of a metal can with a transparent plastic bottom and with sides and top coated with zinc sulfide. Room air is pumped through a fine filter (to remove radioactive dust, but not radon) and into the can; if radon is present, it will emit alpha particles and the can's lining will scintillate.

A sensitive photocell "sees" each flash optically through the transparent plastic, and sends a signal to a counting circuit that adds up all the flashes that occurred over a preset period (we used a five-hour period in our tests). The more flashes, the greater the radon concentration. At the end of the test period, the machine's "memory" stores the total, the counter is reset to zero, and the entire cycle starts again.

A scintillation counter set up in this manner is quite specific for measuring radon: spurious beta and gamma radiation won't be picked up, while the pump-and-filter arrangement keeps nongaseous alpha-emitters out of the system.

2. *Alpha-track devices.* Originally developed at General Electric's Research and Development Center as part of the lunar landing efforts of the 1970s, an alpha-track detector consists of a

small sheet of special plastic, similar to a sunglass lens. Alpha particles striking the plastic cause microscopic pockmarks that are "developed" chemically at the supplier's laboratory for greater visibility; betas and gammas just pass through. To ensure that what is being measured is radon, and not daughters, the plastic sheet is placed in a small canister with a filter across its opening to block solid particles, but permit gases to pass through. After a period of time (typically one month to one year) the detector is placed in an envelope and shipped to a laboratory that examines a portion of the sheet either optically or through computer-assisted electronic imaging, counting pockmarks. The number of pocks is directly proportional to the mean radon concentration "seen" by the device.

A basic characteristic of alpha-track detectors is their insensitivity: a measurement read after, say, three months of exposure will typically consist of fewer than 100 pocks, or "hits." This works both as an advantage—alpha-track detectors are ideal for making long-term measurements—and a disadvantage—they're not well-suited for quick tests.

Alpha-track detectors are relatively expensive: typically, they cost roughly $25 to $50 in quantities of one, dropping to half of this or less for large orders. CU tested two alpha-track devices: the Terradex and the Radtrak.

3. *Charcoal-adsorbent-type detectors* consist of granules of activated charcoal that adsorb gases (including radon) to which they are exposed. When the charcoal has been exposed long enough to become saturated, it is resealed and shipped back to a laboratory where the radioactivity is measured. The level of radon to which the device was exposed is then calculated.

Typically, the charcoal is contained in a metal or plastic canister or jar, with a cap or a tape seal separating it from the ambient air until the detector is opened. As one might suppose, the ratio of gas infiltration rate to the amount of charcoal present determines the period of time before saturation occurs, and hence the time period over which the device "reads" the am-

Terradex

Radtrak

Nodar

Teledyne Isotopes

RTCA

University of Pittsburgh

Air–Chek

bient radon level. Three of the units CU tested—the Nodar, the Teledyne Isotopes, and the RTCA—consist of flat canisters with a mesh across the entire opening. All three of these call for four-day exposures. The DBCA collector is similar in shape, but with a small (about penny-sized) opening in one side, with a multi-layer gauze barrier covering the opening; this device's instructions call for a seven-day exposure. The Air-Chek dispenses with the canister entirely; it consists of an oversized "tea bag" formed of a porous paper. The Air-Chek's instructions call for exposure for up to three days, depending upon humidity.

One of the major motivations for the development of the charcoal-adsorbent detector was the need for a lower-cost test method (the other was to come up with a test that could provide quicker results—alpha-track devices normally won't exhibit enough pocks in short-term exposures to make reliable readings possible). The five models that CU tested ranged in price (for a single unit) from $10 to $50; quantity discounts were also available. If instructions are followed properly, charcoal adsorbent devices can exhibit accuracy as good as, or better than, alpha-track detectors.

How CU Tested

Our first task was to calibrate our laboratory instrument, in order to determine the precise relationship between number of flashes per time period and radon concentration. For this, we were helped by the EPA and the Department of Energy (DOE), who made their "radon room" in New York City available to us. This is a sealed room containing a controlled and monitored level of radon gas.

We were able to check our device's calibration further during the course of the test with a standardizing scintillation cell—one similar to the test cell, but containing a small amount of radon in equilibrium with a bit of radium. Comparable readings of this cell before and after our tests assured us that the counter/timer/storage circuitry of our instrument was functioning properly.

The scintillation counter was then placed in the basement of a house known to have high radon levels (exact levels were not known to CU before we tested, but earlier tests had reported 20–100 pCi/l). Samples of alpha-track and charcoal devices to be tested were placed in the basement, following manufacturers' instructions. At the end of the appropriate test period for each device, CU's engineers—once again following manufacturers' instructions—repacked the testing devices and returned them (with pseudonymic names and addresses) for analysis and reporting. The reported radon levels for each product, compared to the actual levels recorded by our laboratory instrument for that test period, are noted in the Listings. The basis for our Ratings is each device's ability to predict mean radon conditions consistently and reliably over the course of the test program with an accuracy suitable for guiding corrective action.

We anticipated that all the products would, at least, provide reasonably accurate measures of the radon concentration encountered during the time period over which each was tested. In a program announced by the EPA subsequent to the start of this project, all of our devices had passed EPA's test for accuracy. In the EPA's test, samples were exposed in a chamber with a constant level of radon, and sent back to the manufacturer for evaluation. (Manufacturers were not told the level of radon in the chamber). The products CU tested, along with some 200 others, produced results within 25 percent of what they should have in EPA's tests. And, although there were some cases of misreporting during a previous EPA go-round, the industry seems to have settled down to a consistently good level of performance.

EPA's test, however, is a pretty minimal one. A device's ability to measure a constant level of radon accurately may be a very far cry from its ability to predict a long-term average level from a short-term exposure, or to deal with random fluctuations encountered during the test period. CU's tests, and our review of a great deal of data supplied largely by EPA and the University of Pittsburgh, helped us to judge how well the different devices can perform this prediction. Our data aren't anywhere near conclusive, but are illustrative of how and why measurements can go wrong.

What We Found

Our first finding was that all the devices were easy to use, with clear, simple, complete instructions. In each case, the package containing the detector was to be unsealed, placed in the test area for a specified time, then resealed and mailed back to the manufacturer for analysis, along with a form that the householder was to fill out with such information as return address and date and time started and finished. Some forms requested more information, such as where in the house the test took place and other database-type information such as type of heating system, how long the householder had lived in the house, and whether there were smokers in the household, but none proved to be either onerous to fill out or, in CU's judgment, much different from any other in terms of convenience.

Some later samples of Air-Chek that CU purchased illustrated our second finding quite clearly: the radon-detection business is very much in a state of flux. Air-Chek has repackaged its device with a new, streamlined information form complete with a humidity-sensing indicator spot and revised instructions for use: you're supposed to check the spot for color change, and adjust the exposure period from 36 to 72 hours, depending on the color. And CU has been informed by the University of Pittsburgh's Radon Project that, conforming to EPA's prescribed test protocol, newer DBCA collectors will be sent out with a revised instruction sheet. Where the homeowner was formerly instructed to place the charcoal canister in a room "where people spend a lot of time; it should not be in a basement unless a lot of time is spent there," the new instructions call for placing the device in a basement.

Further evidence of the changing industry showed up when we found that some samples of the Terradex detector had mailing labels bearing the Terradex name, but Radtrak's address. When we called Terradex, we were told that the two companies had merged. We were further informed that although devices bearing the Terradex and Radtrak labels were similar in nature, the Radtrak was read using a computerized analyzer, while the Terradex was read optically with a microscope (although this may change in the near future).

We found one minor area of difference in convenience among the models tested. Three models—the Air-Chek, the Teledyne, and the Terradex—did not have return postage prepaid. They required you to provide the two or three stamps needed—one 22¢ and two 17¢, or any combination of others that add up to 39¢ or 56¢—or to take a trip to the post office to get exact postage.

A striking difference we noted between the alpha-track and charcoal-adsorbent types was the time lapse between our mailing in the exposed detectors and our receipt of the results. Reports on the charcoal units were received typically in a week to ten days; the alpha-track devices took a month or more. Taken in conjunction with the alpha-track detectors' longer exposure period, this lengthy processing/reporting time provides a persuasive reason to select a charcoal unit for screening tests, particularly if you're in a hurry for results (see Recommendations).

Some of the reports we received presented data impressively to two decimal points, while others settled for less precise whole numbers. The precision *required* of a radon measuring device probably calls for a brief discussion, as there's a natural tendency to equate more decimal places with virtue. Here goes:

The *accuracy* of a measurement refers to how close it is to the actuality that it's measuring. Its *precision* refers, in informal terms, to the number of decimal places used in expressing the measurement. It's quite possible—and not particularly unusual—to have a measurement with very high precision and very poor accuracy; the reverse is equally possible. There's little point in expressing radon measurements to a high degree of precision. The normal variations in radon concentration over the course of a year mean that a particular measurement over one period of time could easily differ by a factor of one and a half, or two, from a measurement taken at a different time. Further, the uncertainties involved in estimating individual exposures from radon concentrations, or individual risks from exposure estimates, makes extremely precise measurement presentations about as meaningless as expressing your waistline measurement to the nearest thousandth of an inch.

To put things in a better perspective, typical outdoor radon levels are in the 0.1 pCi/l range. The EPA's "target" for indoor radon is 4 pCi/l or less—a level that the EPA estimates will cause 13–50 lung cancer deaths over the lifetimes of each 1,000 people exposed (that represents an increase of 12–47 per 1,000 over the estimates for lung cancer caused by exposure to outdoor background levels). Houses with levels of 20 or more pCi/l are considered candidates for immediate remedial action. The highest level recorded in a dwelling to date was 2,700 pCi/l. Obviously, there's little or no benefit in knowing your house's radon level to the nearest picoCurie/liter, much less to the tenth or hundredth. CU ignored overly precise reporting when we encountered it.

We found that all of the devices we tested provided at least rough approximations of radon levels existing in our test house, and that the ones with revised instructions behaved comparably to their predecessors. All, in the judgment of CU's engineers and consultants, proved adequate for conducting screening or early-warning tests. Some, though, provided results that were closer to the actual radon levels as measured by our laboratory instrument than did others; the alpha-track Radtrak and three charcoal-adsorbent models—the Air-Chek, the Teledyne, and the DBCA collector—all reported radon levels that were consistently within 25 percent of the levels they were exposed to, a high level of accuracy for a test of this type. The RTCA provided us with readings that were consistently 40–50 percent high; the

Nodar reported 25–50 percent low. The alpha-track Terradex gave us readings varying from 5 percent high to 60 percent low.

The accuracy of the detectors themselves, though, represents only a part—and possibly the lesser part—of the accuracy of the entire measurement and risk-assessment process. Any detector can only report on the radon levels it's exposed to, and radon levels in a house can show considerable variation. If the test period is one in which radon levels were higher or lower than average, the device will report higher-than-average or lower-than-average results. As expected, the shorter-term measurements offered a greater likelihood for error (in reporting long-term average levels) than did the longer-term ones. Table 4–A and Figure 4–1 summarize this potential for error, at least for the particular house and time period of CU's tests; the results agree at least approximately with tests conducted by others. It should be noted that the range of readings presented assumes *no* error in the measuring device; it merely reflects the day-to-day variability we observed in the radon level of one house with a radon problem over a period of 85 days during the winter/spring season.

To maximize the reliability of a radon test, it's important to specify all of the particulars of the test period. Some data have suggested that, in several cases, day-to-day and, more significantly, season-to-season variability of radon levels can be large enough to affect seriously the validity of radon measurements. Test performed by the EPA in Butte, Montana, and by Lawrence Berkeley Laboratories in several different locations, suggest that this kind of seasonal variation may be common. Summertime readings of radon in several houses have been reported that indicate it may be possible to obtain measurements as low as one-tenth of the yearly average if three-month testing is done over the summer, with even lower results obtainable with short-term testing. Other researchers—including Dr. Cohen—have not observed such extreme seasonal variations in houses they have monitored. Though it is possible that these very low summertime readings are a local phenomenon or stem from errors in measurement, so few houses have been monitored continuously over a long period of time that "several cases" represents a good chunk of the data available, and it does not appear feasible to assume that any evidence should be discounted. The reasons for this seasonal variation are not clear, by the way. It can't be explained just by an increase in ventilation during warm weather, and theories, although they can provide likely-seeming explanations, don't seem to work in predicting these variations. All in all, CU recommends wintertime testing as more consistent with year-round averages.

TABLE 4–A. POSSIBLE ERRORS DUE TO ESTIMATING AVERAGE EXPOSURE FROM SHORTER-TERM EXPOSURE IN TEST HOUSE (NEW YORK STATE)

Indicated level as percent of actual level

	maximum	minimum
5 hours	891	33
3 days	221	38
4 days	192	70
7 days	148	72
30 days	111	88
60 days	108	98
85 days	100	100

Another problem with estimating radon exposures from tests of radon in a house is choosing a test site within the house. Since radon usually enters a house from the earth below the house, radon levels within a house are likely to be highest in the basement (though not always). Keeping this in mind, and recognizing that most houses will be tested once at most, one measurement procedure—a screening, or "potential," or "early warning" technique—calls for placing the detector in a house's basement, under "closed-house" conditions, to test for radon

FIG. 4–1. VARIATIONS IN RADON LEVELS WITH TIME

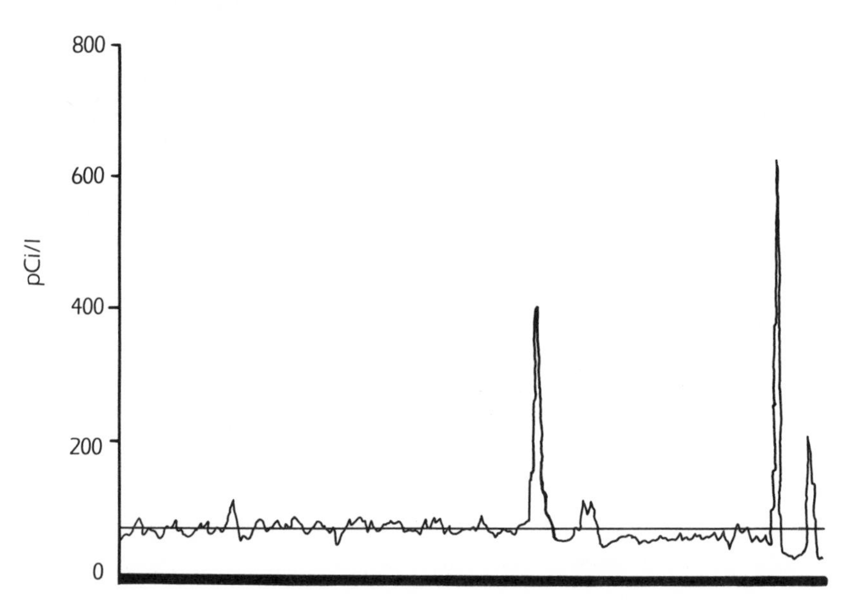

"potential exposure." Quoting from EPA's instructions: "To a reasonable extent, windows and external doors should be closed (except for normal entrance and exit). . . . An external door should not be left open for more than a few minutes. In addition, external-internal air exchange systems (other than a furnace) such as high-volume attic and window fans should not be operating. For measurement periods of three days or less, these conditions should exist for twelve hours prior to beginning the measurement."

Such a procedure will, fairly certainly, result in radon readings that are higher than would be encountered by real people over a year of normal living. (In CU's test house, for example, readings taken in the basement were about twice those in the living room, a relationship similar to those reported elsewhere in the literature.) It will, however, do a reasonably good job of ensuring that the reading will not be misleadingly *low*—unless readings are made in the summer in a house where extremely low summertime radon conditions exist. To help ensure against misleadingly low test data owing to this phenomenon, an effective screening test should be performed in the fall-to-spring period. This procedure will also tend to minimize random variations in measurements caused by wind, weather, and varying family activity.

An alternative philosophy, of which Dr. Cohen of the University of Pittsburgh is a vocal proponent, is to attempt to measure actual radon exposure conditions. To this end, the detector would be placed in whatever room of the house the family spends the most time in, and normal family activity would go on. This procedure has the advantage of providing a more realistic picture than does a closed-house basement measurement. It is more likely, though, to provide mistakenly low readings, and can be expected to provide considerably more variable results than its more rigidly controlled, if less realistic, alternative.

On balance, CU prefers the former approach for the homeowner who wants to determine whether he or she has a radon problem. Of the two types of possible errors, we think that inducing a false sense of anxiety is preferable to lulling one into a false sense of security. Since the consequence of an erroneously high reading in this case would be further tests, the risks of a "false alarm" are minimal. We'll adhere to this concept: an acceptable device used properly, should not provide false reassurances of safety. If this requirement involves the device "crying wolf" on occasion, this possibility should be recognized and accepted.

The EPA, for better or worse, has provided a *de facto* resolution to the disagreement. They have undertaken a large-scale testing and listing of radon testing devices that produced results within 25 percent of what they should have. These are published periodically in their "Radon/Radon Progeny Cumulative

Proficiency Report" (see Recommendations). The EPA has also established test protocols for these devices, which mandate closed-house basement testing. And while there is some pressure for modification of these protocols, at present all of the manufacturers listed in the Proficiency Report are recommending following the EPA protocols—even those who originally suggested other test procedures.

Recommendations

In summation, CU strongly urges that all homes be screened for radon as soon as possible. The most practical approach for most is to use a charcoal-adsorbent detector, under closed-house conditions, preferably in the fall-to-spring period. This will provide the lowest-cost tests and the most rapid, reasonably reliable response, and the earliest warning for that small percentage of the population in the highest-risk situations. For a much broader selection of radon monitors than we were able to test, we can refer you to the EPA's "Radon/Radon Progeny Cumulative Proficiency Report," available at no charge from your state agency (see Appendix A, "State Contacts for Radon").

Once you have received the test results on your house (this usually takes one to two weeks for charcoal-adsorbent detectors), you have to decide what to do about your house. Probably the most practical strategy involves a triage approach, with different plans of action chosen depending upon whether your radon level measurements are high, low, or somewhere in between.

Situations with the highest radon levels—those with readings of 20 pCi/l or more—are those in which accuracy is least important and action is most urgent (*hurry* means *hurry*). These (as a guess) may occur in some 2 percent of residences. Though the closed-house, basement screening procedure almost guarantees that average levels in living areas of homes with these high readings will actually be lower than the reading would indicate, even the lowest actual-living-condition level that's apt to exist poses a risk so large that immediate mitigation procedures are called for.

Houses with readings of 2–5 pCi/l or less (the EPA uses 4 pCi/l as a threshold for recommended action) pose lower risks, though they are not negligible. Such houses are apt to have average radon levels closer to the irreducible outdoor background level, and efforts towards mitigation are apt to be very expensive when measured against the achievable results.

The third case, and the one requiring the most thought, is when the reading obtained from the screening test is between those two limits. In CU's judgment, this case calls for additional follow-up testing, preferably of the areas in which the family spends the greatest amount of time. Radon levels in this intermediate range should be low enough so that the need for correction isn't urgent. We recommend alpha-track detectors in the most heavily used areas of the house, using 3-to-12-month exposures. These longer-term tests provide a more reliable picture of how much radon you're actually being exposed to.

There's nothing holy about our suggested limits—or the EPA's or anyone else's, for that matter. The 4 pCi/l limit proposed by the EPA represents an estimate of a low level that can be achieved in most homes as a practical matter. At that level, the estimated number of lung cancer deaths due to radon exposure per thousand people is 13 to 50—a *much* higher rate than the EPA considers acceptable for *any* other material (one per million is the usual guideline for acceptability). If you wish to be more conservative, pick lower limits and base your actions on them. An exposure to 2 pCi/l will halve the risk of lung cancer (compared to the 4 pCi/L standard); exposure to 8 pCi/l will double it. The urgency for action will depend not only upon the radon levels in your home, but upon the number of hours each day that the house is occupied and upon how long you've lived there. If you and your family are only in the house for half a day, the risk is half that incurred by a family that's almost always in the house. And if you have already been exposed to the radon in your home for, say, ten years, the additional risk of another few months' exposure is minor.

Determining how much risk you're willing to take is an individual concern. The cost of over conservatism is financial; the long-term risk of undue optimism is to life.

Radon in the Water Supply

In some parts of the country—most notably in Maine—fairly high concentrations of radon have been found in houses' private water supplies. This is a problem limited to homes served by private wells: normal municipal water treatment procedures eliminate the radon from the water before it reaches the home. Radon in water is an odd sort of water pollution problem in that much of the radon leaves the water and enters the air as soon as the water leaves the tap.

Radon in water, as a result, is not an ingestion problem, but a contributor to the radon levels in the air. While there is no direct correlation possible between radon-in-water and radon-in-air (it will depend upon the volume of the house and the number of gallons drawn each day), a commonly used estimate is that every 10,000 pCi/l in water will add 1 pCi/l to the air. Typically, well water contains radon at concentrations of less than 1,000 pCi/l, although isolated cases have been found in Maine where concentrations approach 300,000 pCi/l.

Radon from the water supply could be a problem in small enclosed spaces like shower stalls, although the amount of time any one individual spends in the shower or bath is apt to be insignificant; otherwise it generally makes only a small contribution to a house's total radon.

As a health risk, radon from water presents a problem whose seriousness is small with respect to the total radon exposure. (It is, however, extremely *large* compared to other water pollution problems, presenting a nasty problem for the EPA and others whose job it is to regulate permissible water pollutant levels.)

TABLE 4–B. CU-TESTED RADON DETECTORS

CU#	Model Name	Recommended Exposure (days)	Price (one unit)	Comments
The following five models are charcoal-adsorbent types.				
7	Air-Chek	1.5–3	$12	1
6	Nodar	4	$20	2, 3
4	RTCA	4	$30†	3
3	Teledyne Isotope	4	$50	
5	DBCA collector	7	$12	3
The following two models are alpha-track devices.				
1	Radtrak	30–90	$25	3
2	Terradex	30–90	$25	

Key to Comments

1 Has humidity sensor to determine exposure time; more attention required than with others.
2 Careless handling may cause slight charcoal spill.
3 Return postage included.

†indicates shipping is extra

For an additional listing of EPA-listed testing kits, see Appendix B.

Several laboratories, including Terradex Teledyne Isotopes, Radon Testing Corp. of America, and Air-Chek, will test water for radon content; the test costs from $12 to $50.

5

The Uninvited Guest

Major Radon Entry Routes: Slabs and Floors

In the great majority of cases, radon enters houses from the soil underneath or adjacent to its foundations, cellar, or slab. Fig. 5–1 shows a typical construction layout. Most cellars and basements are constructed of poured concrete footers topped by a poured concrete slab and having either concrete or block walls (concrete block or cinder block). After the foundation has been dug and an aggregate base (stone or gravel) put down, the footers are poured and allowed to cure. Once the footer is structurally stable, the basement floor or slab is poured so that its edges will rest on the footer. This slab-footer joint (A in Fig. 5–1), which runs around the entire perimeter of the basement floor, is often not airtight and can act as a conduit for radon infiltration from the ground beneath as the radon seeps through.

An extreme example of this is the use of the "French drain" technique in cellar construction, in which an intentional gap, up to one or two inches wide, is left between the cellar walls and the slab floors to form a drainage path for condensation that may form on the walls. This unintentionally provides a clear pathway for the entry of radon gas.

FIG. 5–1. TYPICAL FOOTER, WALL, AND BASEMENT SLAB LAYOUT, SHOWING TYPICAL ENTRY ROUTES FOR RADON INTO THE BASEMENT

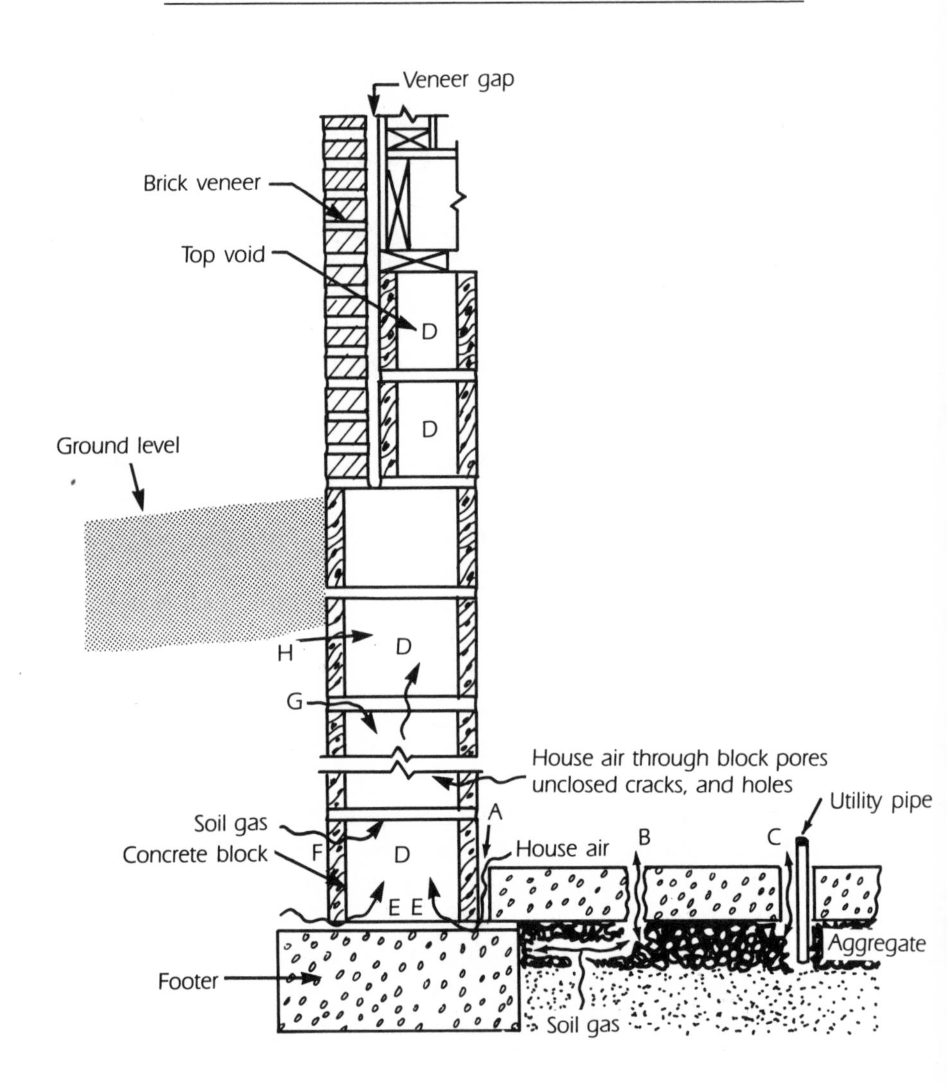

Any cracks in the slab floor (B in Fig. 5–1), or incompletely sealed penetrations of the slab (C in Fig. 5–1), such as floor drains, sumps with exposed soil, toilets, or other piping or utility entry points also act as possible conduits for radon infiltration.

Obviously, exposed-earth cellar floors, earthen root or fruit cellars, and exposed earthen crawl spaces allow for radon to percolate up through the soil and enter your dwelling without restriction. Any such exposed earth beneath your home should be suspected if you find high levels of radon in your living spaces. Be careful not to unintentionally ignore such potential slab-penetration sites as joints or cracks in concrete-floored enclosed porches or additions simply because you cannot gain subterranean access for inspection.

Houses constructed on concrete slabs poured on footers but without cellars or basements also exhibit many of the radon infiltration sites mentioned above. In general, since they have less contact area with the soil, these houses have somewhat lower radon levels than those with basements. However, we have encountered levels over 1,000 pCi/l in a house built on a concrete slab.

Cellar Walls

Radon seeping from soil that is adjacent to cellar walls can and will infiltrate through this apparent barrier. Basement walls are usually constructed of cinder or concrete blocks, which are hollow inside, and those hollow spaces interconnect, forming a single large volume (D in Fig. 5–1) sometimes including the insides of all the basement walls. This volume is often quite high in radon, which can enter through openings in the bottom of the wall (E in Fig. 5–1), through cracks on the outside (F in Fig. 5–1), through grout joints which are often not of the highest quality, or even directly through the sides of the block (H in Fig. 5–1), which are normally relatively thin and porous. Measurements of the radon level inside this volume (D) commonly reveal very high levels. This radon can then enter the basement through unsealed tops of the wall (i.e., at the juncture of the wall and the subflooring that forms the basement ceiling) (J in Fig. 5–1), through unsealed openings in the wall around windows, doors, and pipe and conduit entries, through cracks, or by diffusion through grout joints or the sides of the block.

Poured concrete walls have no hollow spaces inside and are much less porous to radon, and hence present much less of a problem. However, cracks and poorly sealed utility entries can still allow radon to enter the house through them.

Earthen walls in root and fruit cellars offer another, more direct, route for the infiltration of radon into your home.

Stone or rubble cellar walls, more common in older homes that were constructed before more modern block or concrete walling methods became popular, allow more direct radon infiltration from the soil surrounding them. This is especially true if they are not mortared.

Footing Drains

It is not uncommon in some areas to find perforated drain tiles (perforated terra-cotta piping), sometimes called *weeping tiles*, surrounding part or all of the footers of homes to drain moisture away from the foundation. The water collected by these drains is usually routed to a soakaway some distance from the building, as in Fig. 6–3, or directed to a sump in the basement of the structure where it is collected and pumped away for disposal, as in Fig. 6–2. In the latter case, if there is no trap in the line or if the trap is dry, this system serves as an unintended collector for channeling large quantities of radon into the house.

In cases where the drains direct the water to a soakaway removed from the home, the radon gas (from the soakaway, when it is dry) may enter the drains, which then act as an easy, hollow conduit with close access to openings in the footers, walls, or slabs described above. The radioactive gas may enter at footer/slab joints, wall/footer joints, cracks, or other penetrations of the footer, walls, or slab, or even through porous mortar joints between the individual blocks in the walls.

Basement toilets must have a waste pipe going down through the basement floor into the ground, and the seal between this pipe and the floor is rarely tight. The ceramic base of the toilet resting on the floor is not a seal, and it hides the poor seal of the waste pipe from view. This poor seal is a favorite radon entry path.

Fireplaces in the cellar area of a home are likely to structurally abet the infiltration of radon into the building. Completely sealing the small gaps and joints between the block or stone of the basement wall and the stone and brickwork of the cellar fireplace is very difficult. Consequently, poor seals around cellar fireplaces are common.

Utility Penetrations

Into each modern structure, including homes, must come conduits for the utilities that are so often taken for granted. Water, natural gas, and electricity must enter the building at some point, and wastes must leave it. Aesthetics have dictated that as many of these utilities as possible enter below ground, through entries made intentionally in cellar or slab walls and floors. Drains and water supply lines have traditionally been run underground to avoid problems with freezing during the winter months in colder climates.

Increasingly, electric lines and telephone wires are being routed beneath ground for aesthetic reasons. Whatever the justification, underground utility routes often gain access through cellar walls or floor slabs. Frequently, airtight seals have not been employed at these penetration sites, thereby allowing easy pathways for radon entering the basement.

Water Supplies

In some relatively limited geographic areas, most notably in the state of Maine, the groundwater feeding wells used for domestic water supplies pass through rock that is high in radon production. The radon gas dissolves in the water, and is subsequently released into the air inside the house when the water is heated or undergoes the agitation associated with flow through taps and shower heads. Experience indicates that the radon level in indoor air is increased by about 1 pCi/l for every 10,000 pCi/l in the water. Radon levels in well water are typically about 1,000 pCi/l or less, but situations have been found in Maine where levels are up to 300,000 pCi/l, thereby raising the radon level in the air by about 30 pCi/l. If a house using private well water is found to have high radon levels in the air, the water might be tested as a possible source. If it is found to contribute, it is relatively easy to filter the radon out of the water activated charcoal.

Public water systems seldom contribute significantly to elevated radon levels in homes because the radioactive gas usually has plenty of time and opportunity to escape from solution in reservoirs or in aeration and filtration processes.

Natural Gas

When natural gas for heating or cooking in homes is tapped from underground reservoirs or point-of-discovery sites, it may contain radon. Radon is formed by the decay of uranium and radium in rock grains, and it often percolates out of the grains into the pore spaces between them. Natural gas was formed millions of years ago, and also fills the pore spaces of rock in some areas. In addition, the pore spaces in rock are often used as storage reservoirs for natural gas that is pumped into them. In either case, the radon entering these pore spaces mixes with the natural gas and stays mixed with it as the gas is pumped into our homes. When the gas is burned, the radon, of course, is released. Recent studies indicate that natural gas is not a significant source of radon contamination in most areas, but it should not be ruled out if other sources cannot be detected when surveying your home for the cause of elevated radon levels. One clue on this matter is that your neighbors use the same natural gas and should therefore have similar problems if they use it in the same way.

Building Materials

Since all rock and soil contains uranium—2.7 ppm on the average—it also contains its decay product, radium, which serves as a source of radon. Building materials derived from the ground—brick, cement, plaster, clay, ceramics, and so on—therefore are sources of radon. Those materials are generally so tightly formed that the radon gas cannot get out. Thus, with rare exceptions, the structural elements of your home contribute so little radon to the environment that they can be ignored.

Important Contributing Factors

One of the important contributing factors to the ability of radon to build to high levels is the rate at which the air in your home is exchanged with the outdoors. Owing to factors discussed earlier, outside air is very much lower in radon concentration than the air inside affected homes. Exchanging the radon-bearing air inside with the almost radon-free air outside reduces the levels of radon in your home.

Ventilation is the key. Houses that have been tightened in an effort to conserve energy and to reduce heating and cooling costs will exhibit elevated radon levels resulting from the reduction in air exchange. The result is roughly proportional. Such weatherizing actions as weatherstripping, adding storm windows, closing gaps under doors, and caulking around windowpanes may reduce the exchange of air with the outside by about 20 percent. As a consequence of this 20-percent ventilation reduction, a 20-percent increase in the radon level inside the home can be expected.

Pressure is a more important factor. Most homes exhibit a state of slight or partial vacuum compared with the outside environment. Wind blowing past the top of a chimney drags along air molecules from inside the chimney. This provides a pumping action that pulls air up the chimney, creating a partial vacuum inside the house. An aspirator, which employs the flow of water past a tube to suck air out of that tube, is an example of that same principle. Wind blowing against the walls of the house also creates a slight vacuum by a similar process.

Most houses are warmer inside than the air temperature outside, even if only by a few degrees. Since air expands when heated, it becomes less dense and therefore rises. As the warmer air in your home rises and escapes from many exits high on the structure, it "draws" on the air within the building, creating a lowering of air pressure inside.

Even if there are no escape routes, the warmer air inside a house leads to a lower pressure inside than outside. The pressure at any point is determined by the weight of the air above it. Since warm air weighs less than cold air, the weight of air above a point inside a house is less than the weight of air above a point outside, making the pressure inside lower. In most homes, furnaces and hot water heaters use indoor air for combustion, and the hot combustion products escape out a chimney. This effectively pumps air out of the house, creating a partial vacuum inside.

The partial vacuum inside the house caused by any of these processes sucks radon in from the ground, in much the same way that a vacuum cleaner sucks dust up out of a rug. This is the most important factor for transporting radon into a house. Operating a heating source, like a furnace or fireplace, adds to the natural effect.

If your home could somehow be pressurized so that all of the available radon access routes had air flowing from inside the house to outside, then radon could not enter through any of these avenues. Unfortunately, such action is neither feasible nor economically prudent. You will see, however, that pressure can be used to your advantage in defeating radon in certain contexts. This will be discussed in the next chapter.

6

Defeating Radon

If you would like to reduce the radon level in your house, how do you proceed?

There are two approaches to remedial action: reducing the rate at which radon enters the house, or getting it out of the house once it has entered. Reducing the rate at which radon enters your home might be accomplished either by blocking off the pathways through which it enters, or by reversing the direction of gas flow through these pathways from its normal outside-to-inside direction to an inside-to-outside direction, so that the radon does not flow in.

It is only natural to think it would be easy to block radon's entry pathways by sealing cracks, but that is not often the case. Unless there is exposed rock or soil, a sump open to the ground, or some such obvious large area entry path that is easily blocked, it is normally very difficult to obtain significant reductions in radon levels by this approach.

Reversing the direction of gas flow is accomplished by pressure adjustment. Gas always flows from higher pressure to lower pressure; for example, if there is a leak in a balloon or a gas tank under pressure, gas flows out, and air does *not* flow in during the process. Normally a house is under partial vacuum, at a lower pressure than the spaces in the ground immediately adjacent to it. Thus, the gases in those spaces, including radon, flow in—from higher pressure to lower pressure. To reverse this flow, the pressure outside must be reduced to *below* the pressure inside the house by pumping gases from the spaces in the ground to the outdoor atmosphere. When this is accomplished, the radon no longer flows into the house.

The other approach to remedial action is ventilation. Since outdoor air has a much lower concentration of radon (typically 0.1 to 0.15 pCi/l) than indoor air, replacing the latter with the former reduces the radon level indoors. All other things being equal, doubling the rate of air exchange will cut the indoor radon level in half. However, all other things are often *not* equal. For example, a favorite way of increasing ventilation is by using an exhaust fan. This increases the vacuum inside a house and therefore sucks more radon in from the ground. In many cases, turning on an exhaust fan will therefore increase the radon level in a house, even though it greatly increases the air-exchange rate.

Detailed Procedures for Reducing Indoor Radon

Up to this point we have been discussing generalities. The remainder of this chapter deals with detailed procedures. My personal experience in this area is limited, but the EPA has sponsored a great deal of research. I therefore defer to them, and the material presented is largely derived from two of their publications:

Radon Reduction Techniques for Detached Houses—Technical Guidance. United States Environmental Protection Agency, Office of Research and Development, Washington, D.C. 20460, Air Research and Energy Engineering Research Laboratory, Research Triangle Park, N.C. 27711. EPA/625/5–86/019, June 1986.

Reducing Radon in Structures: Training for Radon Diagnosticians, Student Manual. Prepared for: Office of Radiation Programs, Environmental Protection Agency, 401 M Street, S.W., Washington, D.C. 20460. Prepared by: Evaluation Technologies, Inc., 901 South Highland St., Suite 400, Arlington, Va. 22206. EPA Contract 68–01–7030.

A more readily obtainable booklet, distributed free by state public health agencies, is *Radon Reduction Methods: A Homeowner's Guide*, U.S. Environmental Protection Agency Document OPA-86–005, August 1986.

Ventilation of Indoor Radon

Natural ventilation occurs anytime doors or windows are open, allowing for rather free exchange of indoor air with outdoor air. At times, atmospheric conditions will enhance natural ventilation by providing your home with a breeze. Even with all the doors and windows tightly shut, there is still some natural ventilation through entry and exit points sometimes too small to see. Often, the newer and more tightly constructed your home is, the fewer small ventilation routes it has.

Forced ventilation consists mainly of the use of fans to move increased amounts of air into or out of your house compared to the volume of movement that naturally occurs. Forcing outdoor air *in* dilutes indoor radon levels and assists in the outward movement of radon-laden air. Forcing indoor air *out* accomplishes much the same purpose but can lead, unintentionally, to the partial vacuum situation in your house that has already been discussed.

Generally, ventilation as a method for reduction of indoor radon levels is satisfactory only when it does not exact too great a penalty in terms of expense and comfort. Especially during the colder months in cooler climates, extensive ventilation can prove punishing in terms of your family's comfort and the economics of heating your home.

When the source of radon infiltration can be traced to an area of your home with only occasional habitation—the basement, for instance—a cost/benefit compromise can sometimes be struck by blocking off that particular area for periods of disuse and ventilating it vigorously. Be cautious, however, to take appropriate steps to protect pipes from freezing.

In terms of effectiveness, the EPA notes that forced ventilation techniques can be very successful in reducing basement radon levels of "up to 20 pCi/l to levels below 4 pCi/l." The EPA also notes, however, that natural ventilation techniques offer only moderate success rates compared with forced ventilation.

Of course, natural ventilation techniques involve, for the most part, opening your home or the problem area of your home to free exchange with outdoor air. Doors and windows may be opened, crawl spaces vented, or existing vents serviced to ensure their efficient operation, and additional ports to the outdoors may be installed. Unfortunately, in winter this can entail a very high penalty in increased heating costs and very often in comfort.

Ventilation by means of forced air is usually more satisfactory than natural ventilation. This can be done by exhausting air from the basement to outdoors or by blowing air in from outdoors.

The cost of installing natural ventilation, according to the EPA, is minor. At most, it would entail servicing of existing portals to ensure that they remain open. Operating costs can run from little or none in areas where the temperature seldom drops below the comfort zone, to an increase in heating expense of nearly 350 percent in more harsh climates.

Equipment for forced ventilation and its installation is estimated by the EPA to cost no more than $150 (1986 dollars) for fans capable of moving some 240 cubic feet of air per minute. If additional work is required (added utility installation, louvers, etc.) the cost increases proportionately. The EPA estimates the annual operating costs of these fans to run about $100. Again, the heating expense increase can range as high as 350 percent.

How much reduction in radon level can be expected from this ventilation? Let us say your basement has a volume of 10,000 cubic feet (for example, 30 feet long by 33 feet wide by 10 feet high: $30 \times 33 \times 10 = 10{,}000$). If your ventilation system changes 240 cubic feet of air per minute, it changes the air in the basement roughly every 42 minutes ($10{,}000/240 = 42$). That is about 1.5 air changes per hour (ACH). If the ACH without forced ventilation was 1.5, this will double it, cutting the radon level in half. If it was only 0.5 ACH, adding 1.5 ACH will bring the total to 2.0 ACH, and will reduce the radon level by a factor of 4 (2.0/0.5).

From this example, it is clear that the effectiveness of forced ventilation in reducing radon levels increases if the volume of the basement ventilated is small, and if the original ACH is low. If the volume of the basement were only 5,000 cubic feet instead of 10,000 cubic feet, the radon reduction factors would be doubled.

Heat Recovery in Conjunction with Forced Ventilation

If you live in a climate where home heating is required to maintain indoor temperatures within the comfort zone for several months of the year, you might consider installation of a heat-recovery ventilator. This device allows your system to reclaim some of the heat your furnace generates while maintaining the adequate ventilation needed for overcoming your radon problem (Fig. 6–1).

A heat-recovery ventilator—or air-to-air heat exchanger, as the device is often called—typically has a pair of fans to blow stale air out of the house and draw in fresh air. The outgoing warm air and incoming cold air pass on opposite sides of a thin membrane across which heat is easily transferred. In this situation, heat naturally flows from the outgoing warm air to the incoming cold air.

The EPA has noted that "forced air ventilation with heat recovery is a proven technique for reducing indoor air pollutant concentrations in direct relation to the ventilating rates. Heat recoveries of up to 70 percent are possible. Confidence in the effectiveness of this technique should be high."

Costs related to a heat-recovery ventilation system depend upon the size and complexity of the system. EPA estimates range from $400 to $1,500 (1986 dollars) for systems that move

FIG. 6–1.
THREE WAYS TO EXCHANGE HEAT

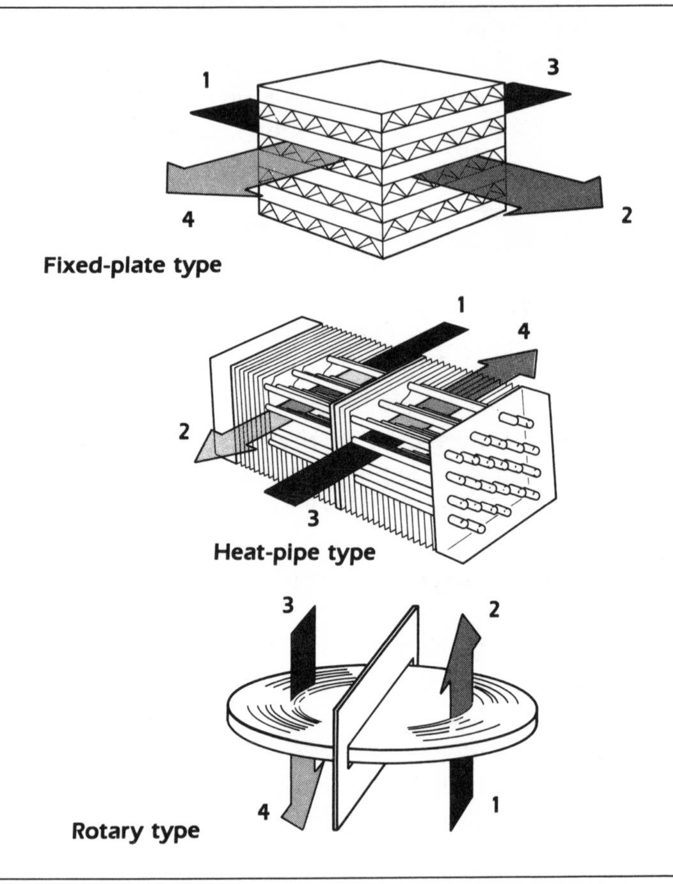

A heat-exchange element is the heart of a heat-recovery ventilator. Fresh outdoor air (1) is warmed as it passes through the exchanger and enters the house (2). Stale indoor air (3) leaving the house is cooled as it transfers heat to the exchanger and is vented outside (4). In the fixed-plate type, heat is transferred through plastic, metal, or paper partitions. The turning wheel of the rotary type picks up heat as it passes through the warm air path and surrenders the heat to the cold air stream half a rotation later. Liquid refrigerant in the pipes of the heat-pipe type evaporates at the warm end and condenses at the cold end, transferring heat to the cold air.

from 25 to 240 cubic feet of air per minute (cfm). The cost of operating the fans in a 240-cfm system would run the $100 estimated above, but the heating cost penalty would drop to perhaps only 40 percent compared with the 350-percent penalty incurred without heat recovery. See the October 1985 issue of *Consumer Reports* for more information about heat-recovery ventilators.

Blocking Radon Entry Routes

Exposed earth or sump systems are obvious targets for radon source control. Sealing or covering such areas with impermeable materials such as aluminum sheet metal is effective. Careful attention should be paid to sealing of any joints that may be needed between existing structural elements and covering materials.

Sealing of known radon infiltration sites is often among the first and least expensive steps taken to deal with low to moderate radon pollution problems in homes. Under ideal conditions this may be all that is required to mitigate marginal situations. Often, it is one step to be used in conjunction with more advanced control techniques.

Sealing of infiltration routes is often limited by your knowledge of these sites and the access you may have to them. Virtually invisible penetrations in your home may exist that you are unaware of, and some may exist that you cannot reach without resorting to extraordinary efforts. Owing to these limitations, the EPA has stated that "a homeowner should not expect sealing of all noticeable cracks or openings to eliminate an indoor radon problem. The potential effectiveness of sealing as a means of significantly reducing indoor radon concentration has been demonstrated in the 30 to 90 percent range. These studies emphasize the uncertainty of successful control with comparable sealing efforts in apparently similar house situations."

Radon infiltration sites of this nature can be sealed by enlarging the opening to a size that permits confident filling with an appropriate caulk or sealant. Your choice of sealants must be compatible with the surrounding material; they must be airtight without any tendency toward shrinkage. Joints in walls and floors are most successfully sealed by a combination of polyurethane membrane sealant and protective cover, or with a nonshrinking grout under a protective concrete cap.

Porous walls, such as those constructed of concrete block or cinder block, as previously discussed, and porous floors can be dealt with effectively by employing interior barrier coatings. Coatings normally used for waterproofing are effective for this purpose. They are typically prepared by mixing two components shortly before application, and they set into a rubberlike sheet that adheres tightly to the surface and is fairly impervious to radon.

Adjusting Pressures to Reverse Direction of Flow

As noted previously, the inside of a house is normally at a lower pressure than the areas in the ground immediately outside the house, which causes the flow of gases, including radon, from outside to inside. This problem is exacerbated when heating devices, fireplaces, cooking appliances, or clothes dryers are in use. Negative pressure can also result from the use of exhaust fans. Whatever its cause, negative pressure in your home can greatly increase the infiltration of radon gas.

One approach to solving this problem is to provide makeup air for combustion appliances. For example, a gas furnace burns by using oxygen from the air and combining it with natural gas to produce heat and exhaust gases that go up the chimney. Removing this air reduces the pressure inside the basement, leading to a negative pressure situation. The solution is to provide air (makeup air) from outside directly to the furnace to participate in the combustion process, thus avoiding the use of the air in the room for combustion. In winter, air coming from outside is cold, but this causes no problem if it is confined to the inside of the furnace and does not enter the room.

According to the EPA, "the American Society of Heating, Refrigerating, and Air-Conditioning Engineers . . . has recommended the provision of outside makeup air for combustion appliances . . . since 1981. They believe that outside makeup air is necessary to ensure the effective and controlled ventilation needed for acceptable indoor air quality."

The EPA also notes that, owing to the seasonal nature of some vacuum-inducing devices and variations in operating conditions and indoor environmental conditions, makeup air systems can result in reducing the negative pressure by as much as 50 percent. The installation consists of a small dampered ductwork system bringing air from outside into the furnace and hot water heater.

Drainage Tile Ventilation

As mentioned in chapter 4, many homes have some sort of perforated drainage tiles, often called "weeping tiles," surrounding some or all of their footers and designed to collect and channel moisture away from their foundations. The water collected in these systems is carried either to a soakaway some distance from the structure or to a sump in the cellar. In the latter case, this system can serve as a hidden highway for the collection and routing of radon gas into your family's living areas.

Sealing of drainage tile systems, or their removal, is both impractical and undesirable—impractical for the expense of excavating such a system, and undesirable because of the valuable water drainage service such a system performs. The easiest and most effective way of dealing with this radon infiltration route is to ventilate it, pumping the air to the outdoors via a small forced-air system, as shown in Fig. 6–2.

Exhausting air from the drainage tile system not only pre-

FIG. 6–2.
DRAINAGE TILE VENTILATION WHERE THE TILES DRAIN INTO A SUMP IN THE BASEMENT

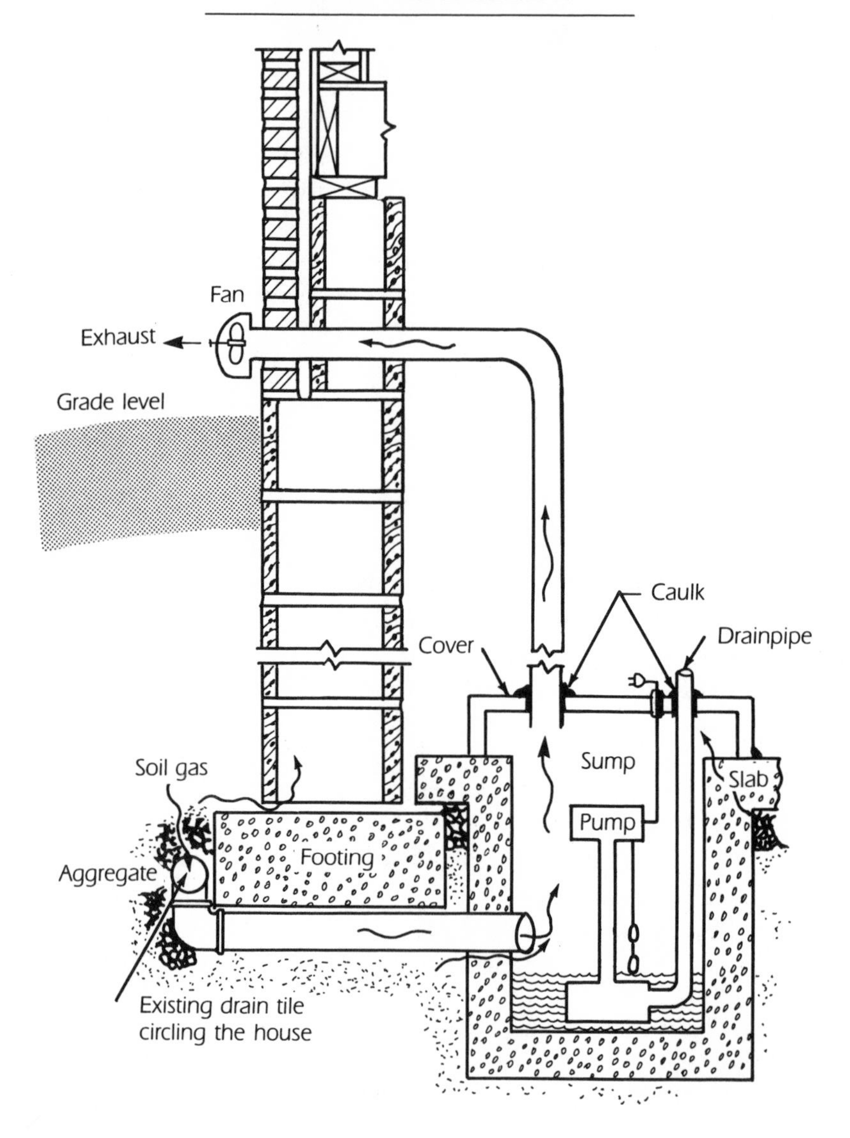

vents the radon in the system from entering the house, but also reduces the pressure in the soil surrounding the house, most desirably to below the pressure inside the house. This second benefit is dependent upon the permeability of the aggregate and soil adjacent to the drainage tiles themselves.

There are limitations to drainage-tile-ventilation effectiveness, however. Some systems are incomplete, and do not surround the entire perimeter of a home, or they may be totally or partially blocked at one or more points. Also, because most drainage tile systems only encompass the exterior perimeter of the footer of a home, ventilation will have little effect on radon emanating from directly beneath the basement or slab.

In studies conducted on the effectiveness of drainage tile ventilation systems, the EPA has found radon reductions of 70 to over 95 percent in homes where drainage tile systems were known to be complete and in good order, and where the ventilation was undertaken in conjunction with the sealing of penetration sites discussed previously.

The pressure-reduction aspect of exhausting air from drainage tile systems can also be effective in cases where the system exhausts into a soakaway rather than into a basement sump, but in that situation, installation requires accessing the system by digging down in the soil outside of the house. An installation is shown in Fig. 6–3. This system consists of a water trap and riser installed in the existing line to the soakaway, along with the fan assembly. The water trap is a necessary feature to ensure that the system will draw air from the drainage tiles and not from the soakaway area.

When connecting the soakaway system, the trap and riser-with-fan must be installed in the line to the soakaway and not in the drainage tile system itself. The location of the discharge line to the soakaway can be estimated by locating the discharge point at the soakaway and then visualizing a straight line back to your house. Dig down until you locate the discharge line at the point where you intend to install the trap and riser. Your selection of installation point will depend upon how far you can or

FIG. 6–3. DRAINAGE TILE VENTILATION WHERE TILES DRAIN TO SOAKAWAY

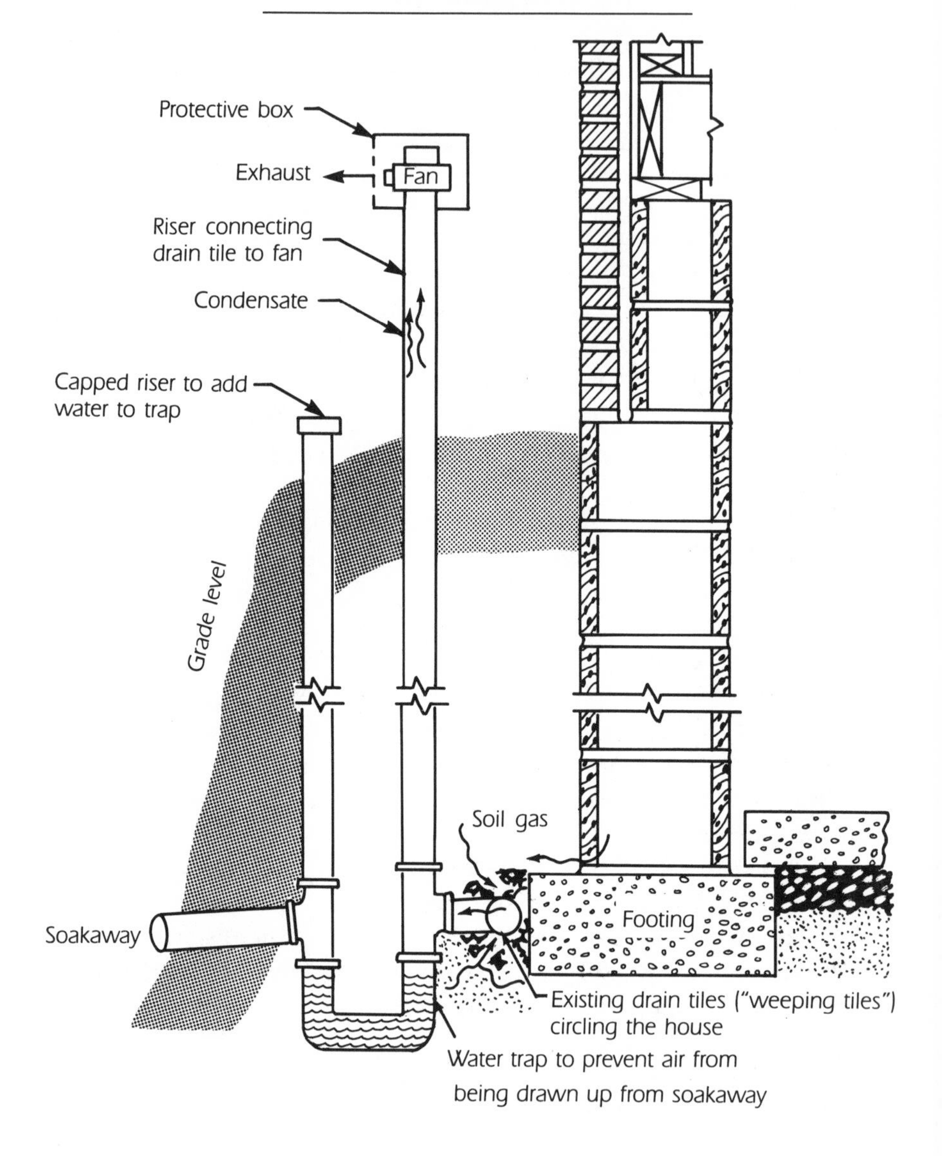

wish to run the electrical line that will supply power to the fan atop the riser, and how near your home you can tolerate the noise of the fan when it is in operation.

The depth at which your discharge line flows to the soakaway will be an added consideration if you live in a colder climate. The water trap you will install must be deep enough so that it does not freeze during the winter months. Such an event would defeat your system and allow radon gas levels once again to build up in your home.

Once you have located your radon discharge point and exposed the discharge line to the soakaway, this line will have to be cut and a section removed sufficient to accommodate installation of a trap-and-riser assembly of four-inch Schedule 40 plastic sewer pipe. Fabricate, or have a contractor fabricate and install, the trap-and-riser assembly as shown in the illustration. Note that a double riser assembly is recommended. This design enables you to periodically check that the trap is, indeed, filled with water and that your exhaust system is working as it should. It is crucial that the trap not run dry and allow your fan to draw air back from the soakaway instead of from the drainage tile system.

The riser on which the fan is installed ***must be on the house side of the trap in order to work as designed.*** Otherwise, your system is useless. This riser should also protrude two to three feet above ground level to allow any condensation that occurs during the winter months to drain into the trap instead of freezing in the fan itself and disabling it. The riser intended for checking and filling the trap need not be elevated above ground level, ***but it should be kept tightly capped*** at all times other than when you are inspecting the trap.

Obviously, all joints in the trap-and-riser system should be sealed airtight, as should be the assembly-to-soakaway-discharge connections.

The fan required by this system, as recommended by the EPA, is a 0.03 horsepower (25 watt) centrifugal fan capable of moving 160 cubic feet of air per minute (cfm) and of drawing up to one inch of water suction before stalling. EPA cost estimates for a fan of this type run from $40 to $100.

Be careful to ensure that the fan *draws air up* from the system and *does not blow air down* into the system. This mistake may drive radon gas into your home under pressure. *The fan itself should be mounted directly on the long riser*. This helps to ensure that it exerts maximum pull on the system and does not waste power overcoming jury-rigged hosing or ductwork. A protective enclosure of some sort should help extend the life of the fan, but should not inhibit exhaust flow. Safety dictates that a protective screen be placed to protect children and animals from the fan's blades. As with the other assembly connections, the mounting of the fan on the long riser must be with an airtight seal; otherwise the fan will simply draw outside air in around the connection and your system will again be defeated.

For drainage tile systems that exhaust into a basement sump, the inclusion of a trap assembly is not required because you are exhausting the sump gas directly to the outdoors. Simply install this system as pictured in Fig. 6–2, but pay careful attention to sealing the sump cover and the fan mounting, as well as any joints in ducting, in order to route the radon successfully outdoors and not into your cellar.

Once your system is installed, be sure to perform routine maintenance on the fan and riser(s), trap, or ductwork. Ensure that your fan is in good working order and that any seals you may have made remain airtight.

If you are able to install a drainage tile ventilation system yourself, the EPA estimates the materials, including the fan, to run in the neighborhood of $300. If you have a contractor undertake the entire project, the EPA cost estimate is $1,200. Contractor costs include the sizable expense of excavation. Preparing the site yourself will reduce the cost of outside help.

The cost of running the recommended fan, the EPA estimates, will be about $15 annually. Because the system will draw some indoor air to the outside, in colder climates the EPA estimates that additional annual heating expenses will be roughly $125.

Ventilation of Hollow Block Walls

Many cellar walls are constructed of concrete blocks or cinder blocks that normally contain hollows. When these blocks are placed one atop another to form the cellar walls, the hollows interconnect to form a single large volume that often contains a very high concentration of radon.

The blocks themselves are made of somewhat porous material, and radon can leak directly through the walls of the blocks and into your basement. Mortar joints between adjoining blocks and the courses of blocks may also allow radon infiltration. The very nature of some of the block material makes them prone to disintegration over time and will allow small penetration points to form. Even the normal settling that a home undergoes over time, and the action of weather, can produce cracks and other radon infiltration points.

These problems can be overcome by pumping air out of the volume inside the blocks, making the pressure inside that volume lower than the pressure in the basement. This will reverse the direction of gas flow, causing it to flow from inside the basement into the blocks, rather than vice versa. In that situation, the radon cannot flow from the hollow volume inside the blocks into the basement, and the source of radon entry is eliminated.

There are limitations to this remedy. Only if the volume is sealed—that is, only if the space inside the block is tightly enclosed—can its pressure be reduced by pumping air out of it. If they are not sealed, you must be able to gain access to the openings in the top course of blocks in order to seal them, and there can be a problem if the bottom course of blocks is not sealed. You must also be able to seal any obvious openings in these walls.

Two methods might be used to exhaust air from these hollows. In the first system, one or two pipes are inserted into the now-sealed top course of blocks in each wall, and joined to a common exhaust to the outside, which is acted upon by a fan.

In the second system, shown in Fig. 6–4, holes are drilled through the bottom of the interior face of the block wall at intervals, allowing for access to the hollow interior. A sheet-metal "baseboard" is run around the perimeter of the cellar and sealed to form a duct circling the cellar and joining all the holes drilled in all the walls. Running an exhaust duct to the outside and installing a drawing fan in that duct will produce a partial vacuum in the walls and effectively draw the accumulated radon to the outdoors before it can enter your living areas.

According to the EPA, the following situations are conducive to ventilation of hollow-block walls:

1. All concrete block walls (including any interior walls that penetrate the floor slab and rest on footings as well as perimeter walls) have a top course with the voids reasonably accessible for being mortared and closed.
2. There is no brick veneer (which may defeat the negative pressure needed to operate the system).
3. There is no fireplace or chimney structure within any block wall.

A baseboard system like that in Fig. 6–4, installed in a home with French drains (see page 155), may have the added benefit of drawing on the sub-slab area and helping to reduce buildups of radon that may be occurring there.

In constructing a system where radon is exhausted via pipes or ducts connected to the walls, every perimeter wall and any interior walls as described above should have at least one exhaust pipe embedded in them. The walls of any heated crawl space must also be ventilated, as should the walls of attached garages, etc.

In EPA tests, one exhaust pipe was placed in each 24-foot length of wall (i.e., in the center, with 12 feet of wall length on

FIG. 6–4. WALL VENTILATION WITH BASEBOARD DUCT

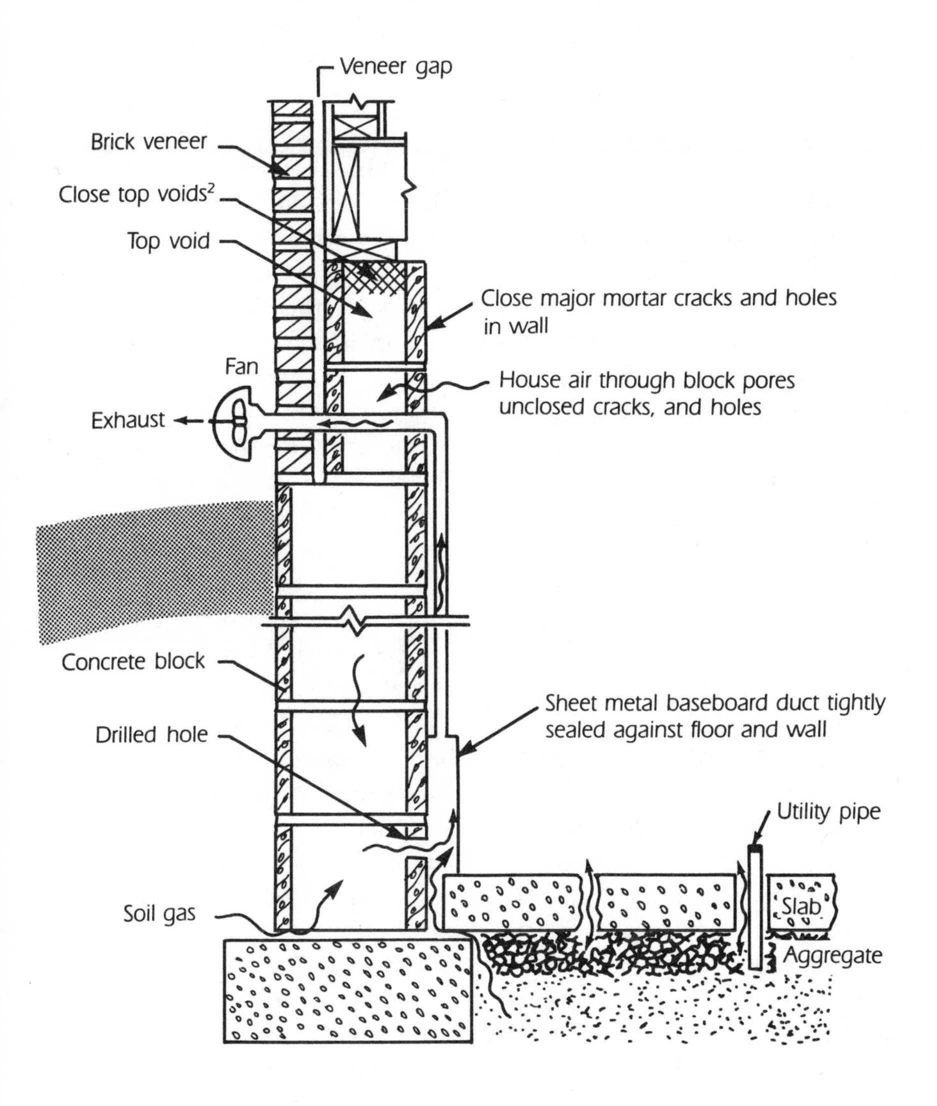

either side). If more than one exhaust penetration is needed in a given wall, EPA guidelines recommend that each penetration be made (assuming two are needed) one-quarter of the way in from the end of each such wall. If the basement of your home is "finished" for use as normal living space, penetrating the cellar walls from the outside and conducting your exhaust system on the outside of these walls may be desirable.

EPA recommendations for sealing the top course of blocks, assuming the crowning row is not of solid blocks, which assure a good seal if intact, include filling each void with crumpled newspaper and then sealing with mortar to a minimum depth of two inches, as shown in Fig. 6–5a. In cases where there is limited access to the top course of blocks, the EPA recommends, again, filling each void with crumpled newspaper, then sealing it with an expanding urethane foam. This material can be extruded through a hose-and-nozzle assembly. Seal with foam to a depth comparable to mortaring. In some cases, urethane foam can also be used for filling the void between the blocks and brick veneers.

When there is no access to the voids in the top course of blocks large enough to permit insertion of crumpled paper or foam delivery devices, the EPA has found success in using the sill plate for sealing, as shown in Fig. 6–5b. Strips of wood were coated on two sides with sealing material and then fastened to cover the gap between the sill plate and the blocks.

In homes with brick veneer that is interconnected with the block-void radon path, access holes were opened to allow insertion of foam sealant between the veneer and sill plate or block walls, as shown in Fig. 6–5c. The access holes were then sealed, and the sill-plate void sealed with coated wood strips as above.

Once the voids are sealed, and the ductwork or piping installed (EPA recommends Schedule 40, four-inch plastic sewer pipe) and sealed, you need only install a 250-cfm fan capable of drawing one-half inch of water suction in your exhaust line or at

FIG. 6–5a. CLOSING TOP VOID WHEN A FAIR AMOUNT OF THE VOID IS EXPOSED

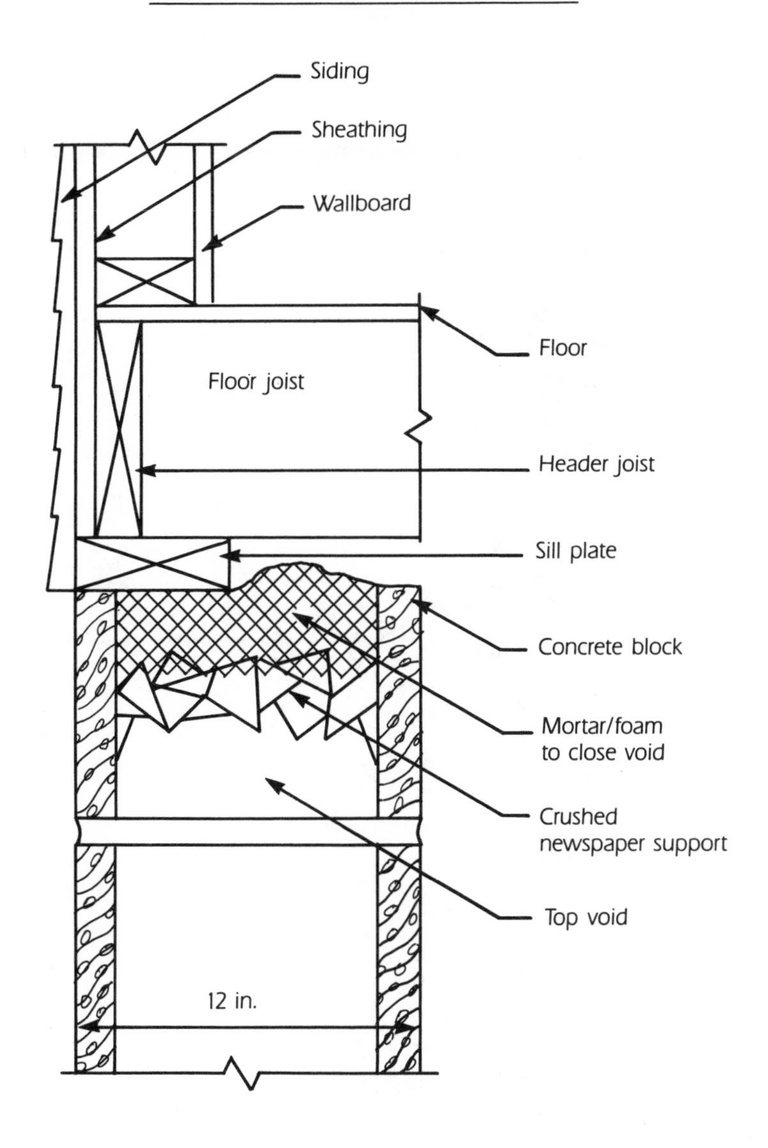

FIG. 6–5b. ONE OPTION FOR CLOSING TOP VOID WHEN LITTLE OF THE VOID IS EXPOSED

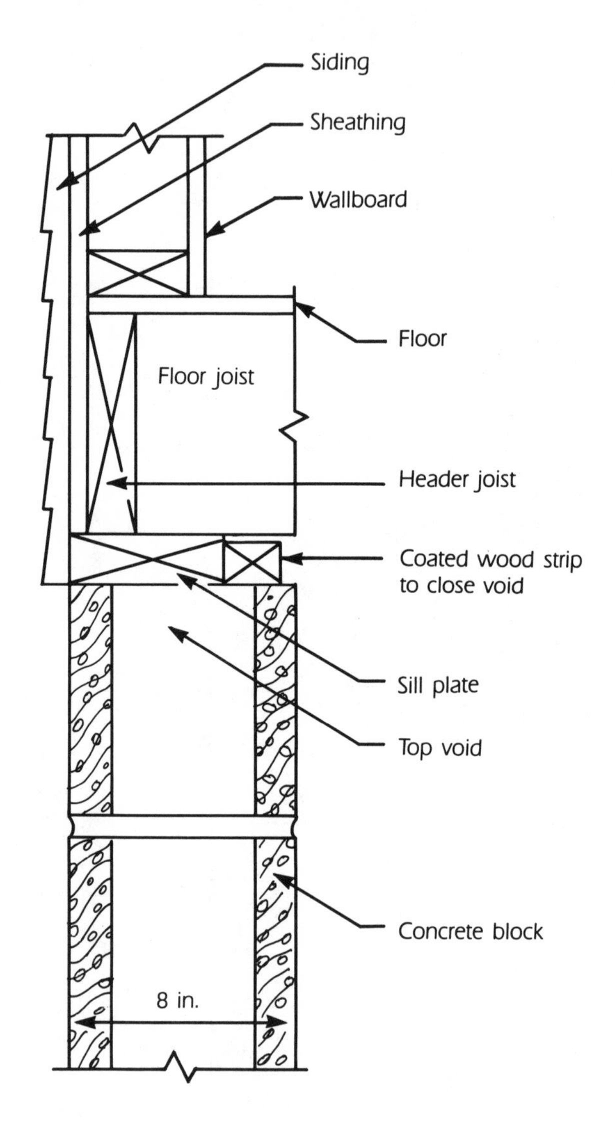

FIG. 6–5c. CLOSING TOP VOID AND VENEER GAP WHEN EXTERIOR BRICK VENEER IS PRESENT

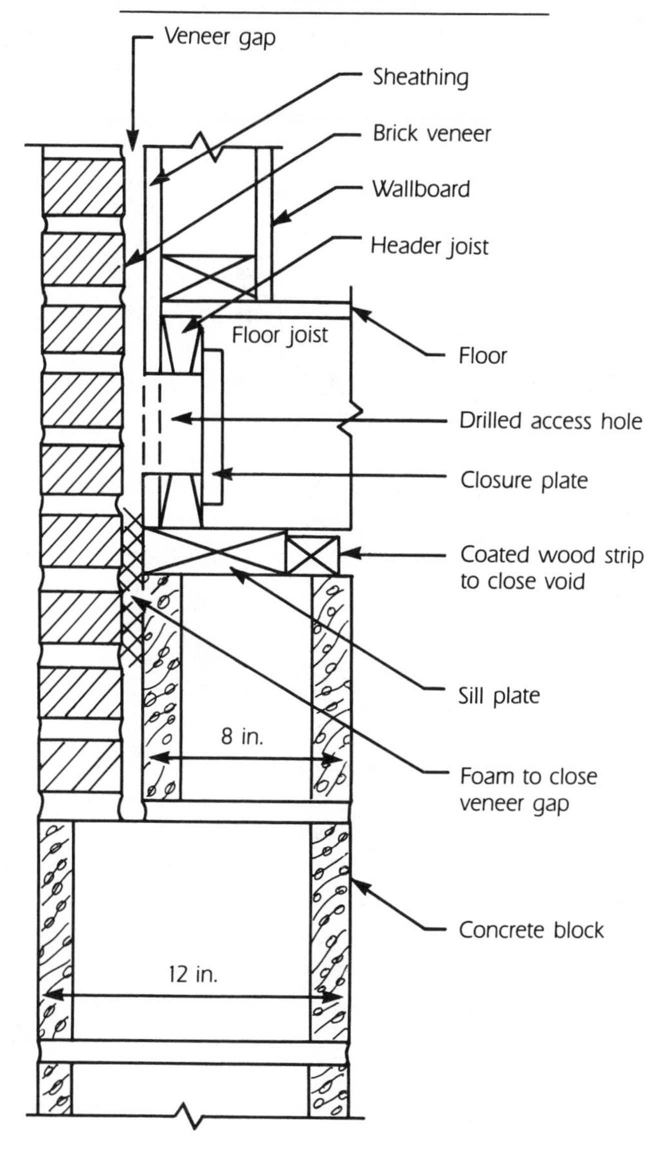

your point of exhaust to the outdoors. Again, be certain that your fan mounts to the system with an airtight seal.

CAUTION: The EPA notes that excessive fan power should be avoided. Excessive suction may cause backdrafting of combustion appliances with the resulting risk of carbon monoxide poisoning.

To test the efficiency of your system, pass a smoke-generating device (such as a smoldering incense stick) over the surface of each wall, as well as along the top and bottom of each. The smoke should be drawn *into* the wall at all points. If it is blown *away* from the wall at any point, this indicates a possible radon infiltration route.

Preparation for baseboard ventilation systems, shown in Fig. 6–4, is similar to that for wall-pipe systems. Instead of installing pipes into wall penetrations of relatively large diameter, however, small drilled holes at the base of the cellar walls provide interior void access in the baseboard systems.

The EPA recommendations include drilling one-half-inch-diameter holes into *each void in each block* in the bottom course of the walls. Baseboard ducts fashioned out of sheet metal are then placed around the perimeter of the cellar, fastened in place with masonry screws and an airtight seal ensured with a bead of asphaltic caulking material. All joints, especially bends, in the ductwork are sealed airtight, including any necessary end caps.

Connect the baseboard system to an exhaust site with plastic sewer pipe (EPA recommends two-inch diameter), and make sure that the entire system is so connected. If the system is not continuous, if it has had to be interrupted for some reason, make sure each section is connected to the exhaust.

Install an exhaust fan (EPA recommends either 250-cfm centrifugal with one-half-inch water draw, or 160-cfm centrifugal with one-inch water draw), as required by your system, to the exhaust and make sure it mounts with an airtight seal. Some baseboard systems require more than one fan. The EPA recommends installing them at opposite sides of the system for most efficient exhausting of radon. *But see the above caution on use of excessive fan power.*

Test your system with a smoke generator as described in the wall-pipe section.

Maintenance of either system consists of routine fan maintenance as recommended by the manufacturer and *careful attention to maintaining the integrity of all seals at all joints, both in-line seals and seals where the system meets existing structural elements.*

The EPA estimates that the cost of a contractor-installed wall-pipe system would be about $2,500. A baseboard ventilation system would cost about $5,000. If you are willing and able to undertake such a task, materials for the installation of either system would run from $100 to $500, according to the EPA.

Operational costs for either system, including running the fan and additional heating costs, are estimated by the EPA to be about $140 per year.

Exhausting Air from Under the Slab

In some cases, radon flows into the house from under the slab beneath your home, whether it is the floor of a cellar or ground base. If the pressure beneath the slab is made lower than the pressure inside the house, this flow direction is reversed and the radon can no longer enter.

According to EPA guidelines, "sub-slab ventilation, by itself, would be most applicable in houses where (1) the concrete slab is expected to contain the major soil gas entry routes (e.g., cracks and other openings), and (2) a reasonably uniform layer of crushed aggregate is known to underlie the entire slab or where soil permeabilities are moderate to high. . . . In concrete block basement houses, the wall-void network probably will always contain major radon entry routes. . . . The use of sub-slab ventilation *in conjunction* with wall ventilation can effectively treat important slab-related entry routes that may not be adequately addressed by wall ventilation. . . ."

The simplest sub-slab ventilation system is shown in Fig. 6–6. It consists of sealing of major slab and wall openings, penetration of the slab itself in two (or more) locations, connection of these penetrations to an exhaust system similar to those described for wall-pipe systems, and an exhaust fan. Pumping air from beneath the slab reverses the pressure gradient, making the pressure lower under the slab than in the basement. The radon cannot then flow from under the slab into the basement.

After preliminary wall and slab sealing has been accomplished, closing all but the smallest penetrations, two holes in the slab should be opened that are large enough to accommodate Schedule 40, four-inch plastic sewer pipe. A small base of aggregate should be laid in a small excavation. Vertical pipe segments are then inserted, the slab penetrations are sealed, and the verticals are connected to an exhaust system with a fan similar to those used in the wall ventilating systems.

EPA guidelines call for placement of the suction points (slab penetrations for vertical pipe members, as in Fig. 6–6) at a rate of one per every 300 to 500 square feet of slab floor area. These guidelines caution that houses with unknown aggregate beneath the slab may require more suction points to deal successfully with the radon problem.

The EPA also suggests placing the vertical suction pipes as near to their ideal locations as possible, but out of primary traffic patterns in order to reduce inconvenience. Placement near vertical load-bearing supports is suggested for this reason.

Two remedies for traffic-pattern interruptions are, first, to place the two suction verticals near cellar walls and run horizontal pipe sections under the slab to your ideal suction sites; or second, to place one vertical suction pipe at the midway point of each outside cellar wall. In all cases, the vertical suction pipes will be connected to an exhaust system.

For installation of any of the above systems, first seal the walls and slab as described earlier. Next, open penetrations in

FIG. 6–6. SUB-SLAB VENTILATION USING INDIVIDUAL SUCTION POINT APPROACH

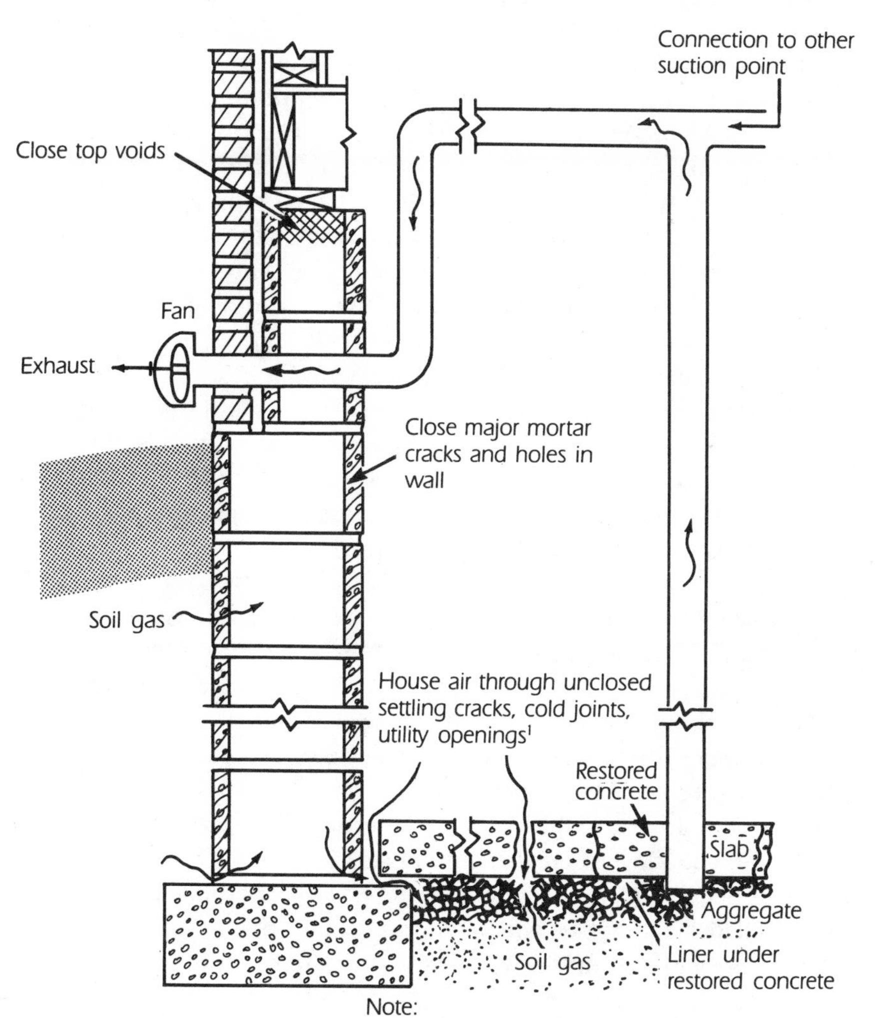

Note:

1. Closing of major slab openings (e.g., major settling cracks, utility penetrations, gaps at the wall/ floor joint) is important.

the slab at the sites you have determined for your suction pipes. You may need a jackhammer to pierce the slab if you are undertaking this project, or this part of the project, yourself. Make the opening large enough to facilitate excavation of a hole beneath each opening that is 1½ feet square and 1 to 2 feet deep. Fill the excavation with gravel until it is level with the underside of the slab. Insert your suction pipe into the gravel until it is at least six inches beneath the bottom of the slab (cf. Fig. 6–6). You will first have to cover the open end of the pipe with a hardware cloth screen to prevent it from being clogged with aggregate.

Once the pipe is in place, cover the aggregate with building felt to keep the aggregate from being clogged with cement, and seal the felt around the pipe with asphaltic caulking material. Make sure that any below-slab joints in the piping are sealed airtight, because they will soon be inaccessible.

If you are running horizontal extensions beneath the slab in order to make placement of your suction verticals more convenient, you will need to break through the slab and excavate passageways to accommodate this variation. Fill the excavations with gravel and prepare the area as above.

Use asphaltic caulking material to attach the building felt to the underside of the slab, and then close the penetrations you have made with high-quality cement. This seal can be enhanced by thoroughly cleaning the broken edges of the slab penetrations and then coating them with an epoxy adhesive. The concrete repair is then completed before the epoxy can set. Level the concrete repairs to match the surrounding slab. After the cement has cured, seal around your vertical suction pipes with asphaltic caulking to ensure a good seal.

Connect your vertical suction pipes to horizontal exhaust pipes of the same size and material, usually placed up in the floor joists, and complete your exhaust system with outlet and fan. Make sure all pipe connections and the fan mounting are sealed airtight.

EPA guidelines suggest the use of a single 250-cfm fan capable of drawing one-half inch of water at low flow, or a 60-to-150-

cfm fan capable of suctions between one and two inches of water.

The system may—and should—be tested periodically, using the smoke test described above for testing the wall ventilation systems. Perform recommended maintenance on the fan in your system and inspect your pipe system periodically to ensure that it is airtight.

Two other sub-slab ventilation systems, which involve running perforated pipe networks either around the entire perimeter of the cellar slab, or forming an interconnected grid under the entire slab, entail far more excavation of existing structures, and therefore seem better suited to situations involving new construction, although they are applicable to existing homes. These will be discussed in the next chapter.

EPA estimates of contractor-installed sub-slab suction ventilation systems, as discussed above, range from $1,000 to $2,500 if no special efforts are required for successful completion of the job. EPA estimates for cost of materials alone, not including the jackhammer, run from $100 to $500 if you choose to undertake a project of this size yourself.

Summary

The various methods of reducing the radon level in a house are summarized in Table 6–A, taken from EPA Document EPA/625/5–86/019 (June 1986), entitled "Radon Reduction Techniques for Detached Houses."

TABLE 6–A. SUMMARY OF RADON REDUCTION TECHNIQUES

Method	Principle of Operation	House Types Applicable	Estimated Annual Avg. Concentration Reduction, %	Confidence in Effectiveness	Operating Conditions and Applicability	Estimated Installation and Annual Operating Costs
Natural ventilation	Air exchange causing replacement and dilution of indoor air with outdoor air by uniformly opening windows and vents	All[a]	90[b]	Moderate	Open windows and air vents uniformly around house Air exchange rates up to 2 ACH may be attained May require energy and comfort penalties and/or loss of living space use	No installation cost Operating costs for additional heating are estimated to range up to 3.4-fold increase from normal (0.25 ACH) ventilation conditions[c]
Forced air ventilation	Air exchange causing replacement and dilution of indoor air with outdoor air by	All	90[b]	Moderate	Continuous operation of a central fan with fresh air makeup, window fans, or	Installation costs range up to $150 Operating costs range up to $100 for fan energy and up to 3.4-fold

	the use of fans located in windows or vent openings				local exhaust fans Forced air ventilation can be used to increase air exchange rates up to 2ACH May require energy and comfort penalties and/or loss of living space use	increase in normal (0.25 ACH) heating energy costs[c]
Forced air ventilation with heat recovery	Air exchange causing replacement and dilution of indoor air with outdoor air by the use of fan-powered ventilation system	All	90[b]	Moderate to high	Continuous operation of units rated at 25–240 cubic feet per minute (cfm) Air exchange increased from 0.25 to 2 ACH In cold climates units can recover up to 70% of heat that would be lost through	Installation costs range from $400 to $1500 for 25–240-cfm units Operating costs range up to $100 for fan energy plus up to 1.4-fold increase in heating costs assuming a 70% efficient heat recovery[c]

Method	Principle of Operation	House Types Applicable	Estimated Annual Avg. Concentration Reduction, %	Confidence in Effectiveness	Operating Conditions and Applicability	Estimated Installation and Annual Operating Costs
					house ventilation without heat recovery	
Active avoidance of house depressurization	Provide clean makeup air to household appliances which exhaust or consume indoor air	All	0–10[e]	Moderate[f]	Provide outside makeup air to appliances such as furnaces, fireplaces, clothes dryers, and room exhaust fans	Installation costs of small dampered ductwork should be minimal Operating benefits may result from using outdoor air for combustion sources
Sealing major radon sources	Use gas-proof barriers to close off and exhaust ventilate sources of soil-gas-borne radon	All	Local exhaust of the source may produce significant house-wide reductions	Extremely case-specific	Areas of major soil-gas entry such as cold rooms, exposed earth, sumps, or basement drains may be sealed and ventilated by exhausting collected air to the outside	Most jobs could be accomplished for less than $100 Operating costs for a small fan would be minimal

Sealing radon entry routes	Use gas-proof sealants to prevent soil-gas-borne radon entry	All	30–90	Extremely case-specific	All noticeable interior cracks, cold joints, openings around services, and pores in basement walls and floors should be sealed with appropriate materials	Installation costs range between $300 and $500
Drain tile soil ventilation	Continuously collect, dilute, and exhaust soil-gas-borne radon from the footing perimeter of houses	BB[a] PCB[a] S[a]	Up to 98	Moderate[g]	Continuous collection of soil-gas-borne radon using a 160-cm fan to exhaust a perimeter drain tile Applicable to houses with a complete perimeter footing level drain tile system and with no interior block walls resting	Installation cost is $1200 by contractor Operating costs are $15 for fan energy and up to $125 for supplemental heating

Method	Principle of Operation	House Types Applicable	Estimated Annual Avg. Concentration Reduction, %	Confidence in Effectiveness	Operating Conditions and Applicability	Estimated Installation and Annual Operating Costs
					on sub-slab footings	
Active ventilation of hollow-block basement walls	Continually collect, dilute, and exhaust soil-gas-borne radon from hollow-block basement walls	BB[a]	Up to 99+	Moderate to high	Continuous collection of soil-gas-borne radon using one 250-cm fan to exhaust all hollow-block perimeter basement walls Baseboard wall collection and exhaust system used in houses with French (channel) drains	Installation costs for a single suction and exhaust point system is $2500 (contractor-installed in unfinished basement) Installation cost for a baseboard wall collection system is $5000 (contractor-installed in unfinished basement) Operating costs are $15 for fan energy and up to $125 for supplemental heating

Sub-slab soil ventilation	Continually collect and exhaust soil-gas-borne radon from the aggregate or soil under the concrete slab	BB[a] PCB[a] S[a]	80–90, as high as 99 in some cases	Moderate to high	Continuous collection of soil-gas-borne radon using one fan (~100 cfm, ≥0.4 in. H_2O suction) to exhaust aggregate or soil under slab For individual suction point approach, roughly one suction point per 500 sq. ft. of slab area Piping network under slab is another approach, might permit adequate ventilation without power-driven fan	Installation cost for individual suction point approach is about $2000 (contractor installed) Installation costs for retrofit sub-slab piping network would be over $5000 (contractor installed) Operating costs are $15 for fan energy (if used) and up to $125 for supplemental heating

[a]BB (block basement) houses with hollow-block (concrete block or cinder block) basement or partial basement, finished or unfinished
PCB (poured concrete basement) houses with full or partial, finished or unfinished poured-concrete walls
C (crawl space) houses built on a crawl space
S (slab, or slab-on-grade) houses build on concrete slabs.

[b]Field studies have validated the calculated effectiveness of fourfold to eightfold increases in air exchange rates to produce up to 90-percent reductions in indoor radon.

[c]Operating costs are ascribed to increases in heating costs based on ventilating at 2 ACH the radon source level: as an example, the basement with (1) no supplementary heating or (2) supplementary heating to the comfort range. It is assumed the basement requires 40 percent of the heating load and if not heated would through leakage still increase whole house energy requirements by 20 percent. Operating costs are based on fan sizes needed to produce up to 2 ACH of a 30 × 30 × 8 ft (7200 cu ft) basement or an eightfold increase in ventilation rate.

[d]Recent radon mitigation studies of 10 inlet/outlet balanced mechanical ventilation systems have reported radon reduction up to 96 percent in basements. These studies indicate air exchange rates were increased from 0.25 to 1.3 ACH.

[e]This estimate assumes that depressurizing appliances (i.e., local exhaust fans, clothes dryers, furnaces, and fireplaces) are used no more than 20 percent of the time over a year. This suggestion that, during the heating season, use of furnaces and fireplaces with provision of makeup air may reduce indoor radon levels by up to 50 percent.

[f]Studies indicate that significant entry of soil-gas-borne radon is induced by pressure differences between the soil and indoor environment. Specific radon entry effects of specific pressurization and depressurization are also dependent on source strengths, soil conditions, the completeness of house sealing against radon, and baseline house ventilation rates.

[g]Ongoing studies indicate that where a house's drain tile collection system is complete (i.e., it goes around the whole house perimeter) and the house has no interior hollow-block walls resting on sub-slab footings, high radon entry reduction can be achieved.

Alternatives for the Future

There are devices currently under development that would directly pump radon out of a house. The most common approach to this device utilizes a system that pumps indoor air through a charcoal column, which removes the radon. Because the charcoal becomes saturated after about 15 to 30 minutes, and cannot, therefore, remove any more radon, outdoor air is brought in at that point and blown through the column to be returned to the outdoors. This flushes the radon out of the charcoal, preparing it to remove more radon from the indoor air. In other words, this system cleans the indoor air half of the time it's in operation, and is being flushed out the other half of the time. While the first system is being flushed, a second, identical system cleans. This second system then flushes while the first system is cleaning. Both systems use the same inlet and outlet ducts; the change between cleaning and flushing is accomplished simply by switching air flows. Devices of this type should be on the market within the next few months.

Another possible future alternative is an air cleaning system that uses ion generators to remove the radon daughters, including those not attached to dust particles, without affecting the radon itself. It has been reported that a combination of an ion generator and a fan can reduce the radiation exposure in a large room by as much as 90 percent. A separate system of this type would be needed for each room in the house. As this book goes to press, the EPA is beginning to evaluate this approach. A system of this type is now on the market for about $400, which would make its use in each room rather expensive. But since ion generators are available for as little as $50, it seems reasonable to expect the price to drop substantially if a large market develops.

7

Buying or Building a Safe Home

The radon level in a house would rarely be an important factor in a decision as to whether to buy or sell that house. In the great majority of cases, the radon level will not be much higher than in other houses in the area. Radon levels can be reduced very substantially at a moderate cost. A typical cost would be about $1,000, and it would rarely be over $5,000. Even the latter would hardly influence a decision to buy or sell.

Nevertheless, in this chapter we offer advice on what can be done to minimize these problems. As has been stressed throughout this book, *the only way of determining the radon levels in a given home or structure, and thus the concomitant health risk to you and your family, is to have the building tested. You cannot rely on area or neighborhood reports of relatively low radon levels in homes to ensure the safety of your prospective home*. The most any report of this kind, such as reports of so-called "radon hotspots," can do is alert you to the enhanced probability of a high radon level in your home, and thus encourage testing of the structure.

If you are buying a house, it would be wise to test for radon *first*. There are several ways to do this. You can ask the seller if he or she has had a test done, and ask to see the results. You can ask the seller to get a test done. You can arrange to get a test done yourself. Or, when you have the home inspected by a structural engineer, you can have that person do it.

There are some pitfalls involved with all of these alternatives, however. A single sample may not give you a reliable reading, because of day-to-day variation, open windows, and so on. Where the detector is placed will affect the reading. And an unscrupulous seller can easily make the reading lower than it should be by taping over the vent in the detector or exposing it for too short a period. Of course, if you trust the seller, you needn't worry about this. But if you don't, there are things you can do.

One is to place the detector in a tamper-proof box. Such a box will give a signal if it has been moved or jostled in any way. An alternative that avoids the possibility of sabotage is to collect a sample of air during a visit and send it to a laboratory for analysis. This requires a special multilayered bag that is impermeable to radon—the gas leaks through paper and most plastics—and a small pump for inflating the bag. Kits for doing this are supplied by some radon testing services.

Of course, you are measuring the radon level only during the few minutes during which the sample is being collected, rather than averaging over a period of several days, which would be more desirable. But the error is unlikely to be more than about 50 percent. This error can be largely eliminated if the process is repeated on two or three occasions. In making such a test, it is important that the house not be "aired out" shortly beforehand.

To the best of our knowledge, there is no such thing as a certified test result. A testing service can only guarantee its results on the assumption that the test was not tampered with. If you don't trust the seller, any measurement results he shows you cannot be relied upon.

There is no *sure* way to avoid radon measurement problems in a house you don't already own, but you can try to include provisions in your contract to protect yourself—requiring, for example, that the seller find and fix any radon problems. You might want to require an escrow deposit for a month or so to guarantee protection if the tests show a radon problem; this approach is being used successfully in New Jersey.

In building a new home, the obvious question is whether the lot can be tested in advance to determine whether radon is a problem. There are relatively simple tests to measure the radon level in soil gas in the ground, to analyze the soil for radium, which is the source of radon, or to measure the rate at which radon is emanating from the surface. Unfortunately, experience has shown that these measurements are very poor predictors of the radon level in a house. It is not easy to understand why this is so. Probably the radon comes up through a rather localized place, and unless you happen to make your measurement at that place, you will not detect it. Even being one foot away could conceivably make a big difference.

There are features that can be added in the construction of a house that will make remedial action much easier and cheaper if the completed house turns out to have a radon problem. In general, they add to the cost of the house, but their cost is much lower than the cost of retrofitting that would be required if these features were not included.

In the following discussion, we will consider a variety of features that can be included in the construction. This will provide the information needed for decision-making.

From the Ground Up

Whether you intend to build your new home with a basement or merely on a grade-level slab, proper preparation of the sub-slab area can be crucial. (For the sake of simplicity, the following information is based upon the assumption of a basement being included in your plans, but much of this information is equally applicable to slab-on-grade construction.)

In addition to the depth-of-cellar excavation normally called for, you might take the soil down an extra few inches, to allow plenty of space for a sub-slab stone or gravel bed. At this point you might consider the installation of a sub-slab perforated-pipe network, mentioned briefly in the last chapter.

Sub-Slab Ventilation Networks

The more extensive, and potentially more effective, sub-slab ventilation network calls for laying a grid of four-inch-diameter, perforated plastic pipe under the entire slab. Begin by laying a single length of *six-inch-diameter* PVC pipe horizontally on top of your aggregate base oriented in the center of your slab area, front to back. This pipe will serve as a manifold for the grid of four-inch perforated pipe to come.

A vertical exhaust pipe of the same diameter should be tapped into the manifold pipe to provide an exhaust route for potential accumulations of radon gas. Make sure the vertical exhaust pipe is (minimally) long enough to clear the slab, once it has been poured, with enough exposed to facilitate joining to a longer exhaust system. If such a system is needed, this vertical pipe may be joined to the vertical pipe illustrated in Fig. 6–6, and the rest of the system may be completed as shown.

Now lay in a grid of four-inch perforated PVC pipe at right angles to the manifold pipe. Cap the outside ends of these smaller pipes and tap them into the manifold pipe. This system virtually ensures effective sub-slab ventilation when connected to an exhaust system as discussed in the last chapter.

The alternative sub-slab ventilation technique consists of laying four-inch perforated PVC pipe around the entire inside perimeter of the footer, about 18 inches in from the inside surface. These pipes are joined at the corner turns with airtight seals and connected with an airtight seal to a vertical four-inch-diameter, *non-perforated* exhaust pipe for later connection to an exhaust system as discussed in the last chapter. Although not as reliable as the grid system, this perimeter ventilation technique often yields good results.

Cellar Walls

Reinforced poured concrete basement walls, minimally four inches thick, are preferable to block walls for radon resistance. *Reinforced* solid block walls are preferable to *reinforced* hollow-block walls, and *reinforced* hollow-block walls are preferable to unreinforced hollow-block walls. Rough stone or rubble cellar walls without thoroughly mortared joints are discouraged. Reinforcing any cellar wall construction for your new home is recommended to minimize subsequent cracking tendencies.

Interior and exterior sealing of all cellar walls with damp-proofing material will minimize diffusion of radon gas through the wall material and any joints in it.

If hollow blocks are used to construct your basement walls, make sure that the top course of blocks is solid, or that the openings in the top course of hollow blocks are thoroughly sealed using the techniques described in the preceding chapter. Also be certain that the bottom course is sealed. These measures would be a good investment in the construction of *every* home, because they are very easy and inexpensive to do during construction, but very difficult and expensive as a retrofit. The hollow spaces inside block walls may be the most important source of radon in American homes. Having the space sealed off at the bottom helps by removing an important entry path of radon from the soil; having it sealed at the top stops an important entryway for radon from the insides of the block into the house; and having both top and bottom sealed makes it easy to exhaust air to the outdoors from the volume inside the blocks if that is necessary. If there is any single measure we recommend in house construction, this is it.

Again, as with slab penetrations, any below-grade cellar wall penetrations should be well sealed to prevent radon infiltration. Penetrations through the wall above ground level are *not* a radon problem. Radon levels in soil gas are much higher than those inside houses (typically 100 pCi/l), whereas radon levels in outdoor air are much lower than those inside houses (typically, 0.15 pCi/l).

Chimneys or flues that involve penetration of cellar walls, and especially such unnecessaries as cellar fireplaces, are best not included in your home design from the standpoint of radon. It is difficult to achieve adequate radon-resistant seals between such structures and cellar walls. One alternative is a free-standing fireplace design that can be safely vented without providing radon with a potential infiltration site. Remember, too, that fireplaces and their chimneys contribute to the undesirable negative pressure (partial vacuum) situation in your new home.

Central Air-Conditioning Systems

If your home design incorporates central air conditioning, it is best if it *does not* run condensation drains directly to the sub-slab stone bed. Utilize the techniques already described for slab-penetrating floor drains or, better still, drain condensation from central air-conditioning systems to daylight or the sewer system.

Exhaust Systems

When considering interior exhaust systems (including, but not limited to, bath fans, dryer exhausts, range fans, kitchen fans, whole-house fans, etc.), beware of adding unnecessarily to the undesirable negative pressure situation discussed in the previous chapter. It would be better if such exhaust systems incorporated fresh-air supplies to offset vacuum tendencies.

Either of these sub-slab ventilation systems can be vented, eventually, to the outside at one of two exit points. The first is a point of penetration in a cellar wall that is above ground level, or in a dug-out well like those commonly used outside basement windows. The second possible exit point calls for running exhaust pipe vertically to a point one and one-half to four feet above the roof of your new home. If this second route is chosen, six-inch-diameter PVC exhaust pipe should be used for the vertical in order to reduce the pressure drop associated with this increased length. You may find that this through-the-roof option provides enough passive ventilation due to the natural chimney effect, aided, perhaps, by a wind-driven turbine blower. Alternatively, you might incorporate provision for an electric exhaust-fan installation, should this prove necessary.

Care should, of course, be taken to ensure airtight seals at slab penetration points.

The Slab

Once the footers are in, and any sub-slab ventilation system you may choose has been installed, attention must be paid to the process of laying the slab itself.

Installation of a sub-slab vapor barrier has become an integral part of modern construction techniques. It has also been found to have value in combating the entry of radon gas if installed in the following manner.

If you have chosen to incorporate a sub-slab ventilation system into the construction of your home, before you prepare to install the vapor barrier prior to slab pouring, you must fill in around and above this system with sub-slab aggregate material to a suitable depth relative to your footer. In order to protect the vapor barrier from damage by contact with the rough aggregate material, next put down a layer of sand on top of the aggregate.

Use 24-mil vapor barrier material, *sealed at the seams*, on top of the sand, and then cover the vapor barrier with a protective material (e.g., building felt) to protect it from accidental punctures by workmen who may come in contact with it.

Once the vapor barrier and its protective material are in place, it is time to pour the slab. If at all possible, the slab should be installed in a single pour. If this is not possible, use as few pours as possible to minimize the number of joints. Reinforce the slab with wire mesh to help prevent large cracks associated with age and house settling. Pour the slab to a minimum thickness of four inches, right up to the walls of the basement, and seal the perimeter crack with long-lasting, nonshrinking caulking material (e.g., polyurethane-based caulk). As an additional precaution, you may wish to coat the floor surface of the slab with commercial, high-grade, damp-proofing sealant.

Slab Penetrations

If floor drains are to be incorporated into your house design, it is best to incorporate water traps in the piping and ensure that these systems drain to daylight or, preferably, to interior or exterior drains or municipal sewer systems. Be certain to seal completely around any drains at points of penetration of your vapor barrier and, most important, the slab itself. Always use solid-wall piping for floor drains, and seal any joints thoroughly.

Avoid the use of interior sump systems. If you must use an interior sump, be sure that it is tightly sealed. It should be designed so as to make exhaust ventilation easy to add if it is needed. Do not consider using any exposed-earth sump.

Any slab penetrations for utilities should incorporate airtight seals.

Combustion Appliances

Combustion appliances such as cooking stoves, ranges, and furnaces may be major contributors to the negative pressure (partial vacuum) environment in your home. They take air out of the room, combine it with gas or oil to release heat and produce exhaust gases (mostly carbon dioxide and water vapor) and send these exhaust gases up the chimney and out of the house. Effectively, then, they are removing air from the room, which creates a partial vacuum. As with the exhaust systems mentioned above, it is preferable for the design of your new home to incorporate adequate fresh-air supplies for these appliances. That is, it is better to bring fresh air in directly from outside to participate in the combustion process than to use room air.

Typically, the fresh air comes in through a four-inch duct. Since it does not mix with the room air, it does not affect the temperature in the room, and hence does not affect heating costs. Consult your home designer/builder for designs appropriate to your particular situation.

Air-Filtration Systems

Air-filtration systems that are designed to remove dust from the air in your new home are of dubious value as contributors to a radon-resistant home. Since most radon daughters are attached to dust particles, such systems do greatly reduce the total number of radon daughters, but by removing the dust they increase the numbers of radon daughters that remain unattached. Since these unattached radon daughters are much more hazardous to your health than attached daughters, air-filtration systems do not reduce your health risk from radon. Of course, these systems may have other benefits unrelated to the radon problem.

Since radon is a gas, it is unaffected by filtration or by any other commonly used air-cleaning system, such as electrostatic precipitators or ion generators.

Water Supplies

Although the most troublesome source of radon gas in terms of case studies is the soil beneath and surrounding your home, your water supply, if the source is a private well, may represent a significant route of entry.

There is no need to consider this problem during construction. Once your well water becomes available, its radon levels can be determined using methods discussed elsewhere in this book. If testing results indicate that your water supply represents a significant source of radon infiltration, there are two viable methods of controlling the problem.

Granular activated carbon (GAC) adsorption/decay utilizes a bed containing 1.5 to 3 cubic feet of activated charcoal to remove up to more than 99 percent of radon contamination in the water used by a typical household.

Aeration has been used successfully in the reduction of radon levels in individual-home water supplies by the *diffused bubble method* and the *spray method*. The Division of Health Engineering of the State of Maine Department of Human Services has devised a spray aeration device that removes up to 93 percent of the radon from a well tested to contain very high radon levels. Private entities in that same state have developed a radon-reduction method for well water based on diffused bubble technology. Aeration involves noise levels that some homeowners may find objectionable.

Because of the relatively limited demand for water-treatment systems generally, and because of their limited sources of development, if your well is found to contain dangerous levels of radon gas, you might contact the Maine Department of Human Services Division of Health Engineering, or your local Environmental Protection Agency office, for advice tailored to your individual problem.

8

Questions and Answers

I get several telephone calls and several letters every day from people with questions about radon. In this chapter, I give some of the questions most frequently asked and their answers.

Q My husband has had headaches (or upset stomach, or back pains) lately. Could this be due to radon?

A No. The only health effect of radon, known or expected, it that it can cause lung cancer.

Q We have a strange odor in our basement. Could that be due to radon?

A No. Radon has no odor and can do nothing that might cause an odor.

Q The paint on furniture in our basement is discolored. Could this be due to radon?

A No. Radon can have no chemical effects. It is chemically inert.

Q Since shortly after we moved into our new house, I have had a cough. Could it be due to radon in this house?

A No. Even if the cough is due to incipient lung cancer, that lung cancer could not be due to the radon in the new house because it takes a minimum of ten years after exposure for lung cancer induced by radon to develop.

Q I just heard about radon for the first time, and know very little about it. Should I get a measurement of the radon level in my home?

A If you worry about radiation in any way, shape, or form—nuclear power, bomb test fallout, medical X rays, or anything else—you should get a measurement, because nearly all Americans get far more radiation exposure from radon in their houses than they get from any of those sources. Even a small reduction in the radon level in your house can far more than compensate for those other exposures.

Q Is radon a new problem?

A No. It has been a problem since people have lived in houses. When they lived in caves, it was far worse.

Q Has the recent emphasis on tightening houses to save energy caused the radon problem?

A It has increased radon levels only by about 10 percent, which is a relatively trivial change.

Q Is radon mainly a problem in new houses?

A No. Houses over 100 years old have radon levels similar to those in new houses.

Q Why are we just learning about radon problems now? Were they just recently discovered?

A They have been recognized by scientists since the mid-1970s, but the media has been featuring the problem much more prominently since about 1985.

Q Should I worry about the radon in my child's school?

A Schools and other public buildings generally have much lower radon levels than houses, probably because they are larger, better ventilated, and have more tightly sealed basements. Also, your child spends more time at home than at school. Radon in schools should therefore be of much less concern than radon in your house.

Q Should I worry about radon in my workplace?

A Probably for the reasons stated above, stores, office buildings, and other public buildings generally have much lower radon levels than houses do, and you probably spend much more time in your house. Therefore your home should be your first concern related to radon.

Q My neighbor has a high radon level in his house. Does that mean that I have a radon problem?

A Not necessarily, but it is more likely than if your neighbor had a low radon level.

Q Is radon more dangerous for elderly people?

A No. In fact, it is *less* dangerous because it takes 10–50 years after exposure for the lung cancer to develop, and this would usually exceed the expected life of an elderly person.

Q Is radon more dangerous for young children than for their parents?

A There are differences in lung geometry and breathing rate. When these are considered, it is somewhat more of a risk to children, but less than twice as much. In addition, the incidence of lung cancer in humans is extremely low up to age 35, and some believe that the effects of radiation on young children are reduced by this long waiting period.

Q I live near a nuclear power plant. Does that affect my risk from radon?

A No. The effects of the two are simply additive. Also, living near a nuclear power plant typically gives you less than 1 percent as much radiation as you would get from radon if your house had average radon levels.

Q I live near a toxic waste dump; does that effect my radon problem?

A No. The two problems are unrelated.

Q I smoke cigarettes. Does that affect my radon problem?

A Yes. Lung cancers resulting from radon appear at a younger age in smokers, and smoking may also increase your total risk from radon exposure. On the other hand, if you are willing to accept the risks of smoking, the risks of radon are probably trivial by comparison.

Q My house has a high radon level. Should I sell it?

A No. In nearly all cases, the cost of fixing the radon problem is much lower than the cost of selling a house and moving. Also, to sell it without informing the buyer is unethical. Many states are now considering laws that would make a radon measurement a requirement in all real-estate transactions.

Q How can I get the radon level in my house tested?

A Contact your state or local public health authorities, or the Environmental Protection Agency Office of Radiation Programs for recommendations. A large number of firms offer measurements. Also see chapter 4 of this book, which gives the CU recommendations.

Q When is the best time to get a measurement of the radon level in my home?

A The windows should be closed during the measurement. If this is not convenient in summer, the measurement should probably be delayed. If the windows can be kept closed, there is no reason why a summer measurement would not be indicative of your radon problems. It should be recognized that radon levels are generally about 60 percent higher in winter than in summer, but since the annual average exposure is the important information, a winter measurement is also unrepresentative. In either case, it is a simple matter to estimate the annual average.

Q What is the best type of measurement to get?

A The best measurement is a full-year test with a nuclear track detector, since it averages over seasonal variations. If you are not willing to wait that long, a measurement that averages over at least a few days is most desirable. A measurement that averages over less than one day is less reliable, but would give a good first estimate of your problems in most cases.

Q Is a mail-order, do-it-yourself measurement as good as one where a person comes to your house?

A It is normally every bit as good. Instructions for do-it-yourself measurements are exceedingly simple, and very little can go wrong.

Q Is it worth paying more for a measurement to get prompter service?

A Not ordinarily. There is no great rush to find out about the radon level in your home. Radon does harm by exposing people over many years. A few days more or less of exposure can make little difference.

Q In what room of the house should the measurement be made?

A I differ with EPA recommendations on this. The EPA recommends measurement in the basement because, if the basement level is low, it is fairly sure to be low throughout the house. I dislike this idea because it can overestimate your exposure by about 200 percent. I feel that the important information is your exposure, since that is what determines your risk. I therefore recommend a measurement in the main living area of the house, where you and your family spend your time. Since radon levels are fairly similar throughout the living areas, it doesn't matter much where you make the measurement.

Q How many measurements should be made?

A In most cases, one measurement is enough. If you spend a lot of time in the basement, you may want an additional measurement there. If you keep your bedroom closed off from the rest of the house, you may want an additional measurement there. But in normal situations, a single measurement in the main living area of the house will give you a good estimate of your radon problems.

Q How accurate is the measurement?

A In nearly all cases, accuracy is not a problem. If your level is very low, accuracy may be poor, perhaps within 50 percent, but in that case it rarely matters. If your radon level is above 2 pCi/l, the uncertainties would normally be no more than 30 percent, and 15 percent is more typical. It is difficult to imagine why that accuracy would not be adequate. Estimates of risks from radon are uncertain by about 50 percent, so *precise* measurement is of little value.

Q The radon level in my house is x pCi/l. What is my risk?

A Living your whole live in that house, spending 75 percent of your time indoors, would give you x chances in 300 of dying from radon exposure. Put another way, this would reduce your life expectancy by x times 25 days. Compare that with other risks in Table 2–A, page 24.

Q The radon level in my house is x pCi/l. Is that a safe level?

A It is impossible to draw a line between what is safe and what is unsafe. Your risk is proportional to your exposure, at any level. The relevant question is whether you find the risk acceptable, or if you want to take action to reduce it. The information for making this decision is presented in chapter 2. If you have read the discussion there, you know as much as you need to know about the size of the risk, and only you know your personal value system. You should decide. For purposes of simplification, the EPA has arbitrarily set 4 pCi/l as the acceptable level. But you may decide that a lower or higher level of risk is acceptable to you.

Q Should I get a second measurement?

A If you are considering expensive remedial action as a result of your first measurement, you should get a second measurement to confirm it. While large errors are highly unlikely, they are always possible. If the results of your first measurement are such that you are not considering remedial action, I advise letting the matter drop.

Q The result of my measurement convinces me that urgent action is called for. What should I do?

A Consult chapter 4 of this book, and perhaps also the EPA reports cited there. You might also contact your state or local health officials for advice. They can give you a list of contractors who offer remedial action services. In general, these contractors do *not* guarantee to reduce the radon level in your home to a specific level for a definite price, because it is very difficult to predict the effectiveness of remedial action in advance. If you can find a contractor who will guarantee a definite radon reduction, that guarantee is very valuable.

Q I am planning to build a house. Can I test the lot to determine whether I will have a radon problem when the house is completed?

A You can obtain measurements for the concentration of radium (the source of radon) in soil, for the concentration of radon in soil gas, and for the rate at which radon is emanating from the soil at any one point. However, our studies have found that there is very little correlation between the results of these measurements made just outside the house and the radon level inside the house. Perhaps the explanation is that the path for radon into a house is a very localized one, and unless your test happens to be at the right spot, no indication is obtained.

Q In building my house, is there anything that can be done to make remedial action easier if it should be required?

A These questions are discussed in chapter 5. One thing that would be relatively cheap and very useful is to seal the tops and bottoms of basement block walls. A more expensive, but often effective, measure is to lay a system of perforated pipe in the gravel bed under the basement (or concrete slab) floor that can easily be pumped out from inside the house.

Q Should I get a test for the amount of radon in our water supply?

A The only harmful effect of radon in the water supply is that it may increase the amount of radon in the air. The radon in the air should therefore be measured first. Only if this is unacceptably high might it be useful to test the water. A water test would be worthwhile only if the water comes from a well.

Q How can I get a test of the amount of radon in my water supply?

A Tests are commercially available. Recommendations can be obtained from public health officials.

Q If a test of the water indicates that it may be an important contributor to the radon level in the house, what can be done about it?

A It is relatively easy to install a filter that will remove the radon from the water. Other measures are discussed in Chapter 7.

Q How can I get further information?

A Call your state or local public health department. Some states have "800" telephone numbers. See Appendix A.

Appendix A

State Contacts For Radon

Alabama	James McNees Radiological Health Branch Alabama Department of Public Health State Office Building Montgomery, AL 36130 205-261-5313
Alaska	Sidney Heidersdorf Alaska Department of Health and Social Services P.O. Box H-06F Juneau, AK 99811-0613 907-465-3019
Arizona	Paul Weeden, Arizona Radiation Regulatory Agency 4814 South 40th Street Phoenix, AZ 85040 602-255-4845
Arkansas	Division of Radiation Control and Emergency Management Arkansas Department of Health 4815 W. Markham Street

	Little Rock, AR 72205-3867 501-661-2301
California	Steve Hayward California State Division of Laboratories 2151 Berkeley Way Berkeley, CA 94704 415-540-2134
Colorado	Martin Hanrahan Radiation Control Division Colorado Department of Health 4210 East 11th Avenue Denver, CO 80220 303-320-8333, Ext. 6246
Connecticut	Laurie Grokey Connecticut Department of Health Services Toxic Hazards Section 150 Washington Street Hartford, CT 06106 203-566-8167
Delaware	Division of Public Health Delaware Bureau of Environmental P.O. Box 637 Dover, DE 19901 302-736-4731
District of Columbia	Veronica Singh DC Department of Consumer and Regulatory Affairs 614 H Street, NW, Room 1014 Washington, DC 20001 202-727-7728
Florida	Harlan Keaton Florida Office of Radiation Control

	Building 18, Sunland Center P.O. Box 15490 Orlando, FL 32858 305-297-2095
Georgia	Tom Hill Georgia Radiological Health Section 878 Peachtree NE, Suite 600 Atlanta, GA 30309 404-894-5795
Hawaii	Env. Protection and Health Services Division Hawaii Department of Health 591 Ala Moana Boulevard Honolulu, HI 96813 808-548-4383
Idaho	Larry Boschult Radiation Control Section Idaho Dept. of Health and Welfare Statehouse Mall Boise, ID 83720 208-334-5879
Illinois	Illinois Department of Nuclear Energy Office of Environmental Safety 1035 Outer Park Drive Springfield, IL 62704 217-546-8100 800-225-1245 (in state only)
Indiana	David Nauth Division of Industrial Hygiene and Radiological Health Indiana State Board of Health 1330 W. Michigan Street, P.O. B Indianapolis, IN 42606-1964 317-633-0153

Iowa	Bruce Hokel Bureau of Environmental Control Radiological Health Section Iowa Department of Public Health Lucas State Office Building Des Moines, IA 50319-0075 515-281-7007
Kansas	Harold Spiker Kansas Department of Health and Environment Forbes Field, Building 321 Topeka, KS 66620-0110 913-862-9360 Ext. 286
Kentucky	Donald R. Hughes Radiation Control Branch Kentucky Department of Health Services 275 East Main Street Frankfort, KY 40621 502-564-3700
Louisiana	Jay Mason Louisiana Nuclear Energy Division P.O. Box 14690 Baton Rouge, LA 70898-4690 504-925-4518
Maine	Gene Moreau Division of Health Engineering Maine Department of Human Service State House Station 10 Augusta, ME 04333 207-289-3826
Maryland	Roland Fletcher Maryland Department of Health and Mental Hygiene 201 Preston Street, 7th Floor Mailroom

	Baltimore, MD 21201 301-333-3130
Massachusetts	Bill Bell Radiation Control Program Massachusetts Department of Public Health 23 Service Center North Hampton, MA 01060 413-586-7525 or 617-727-6214
Michigan	Robert DeHaan Michigan Department of Public Health Division of Radiological Health 3500 North Logan, P.O. Box 30035 Lansing, MI 48909 517-335-8193
Minnesota	Bruce Denney Section of Radiological Control Minnesota Department of Health P.O. Box 9441 717 SE Delaware Street Minneapolis, MN 55440 612-623-5350
Mississippi	Gregg Dempsey Division of Radiological Health Mississippi Department of Health P.O. Box 1700 Jackson, MS 39215-1700 601-354-6657
Missouri	Kenneth V. Miller Bureau of Radiological Health Missouri Department of Health 1730 E. Elm P.O. Box 570

	Jefferson City, MO 65102 314-751-6083
Montana	Larry L. Lloyd Occupational Health Bureau Montana Department of Health and Environmental Sciences Cogswell Building A113 Helena, MT 59620 406-444-3671
Nebraska	Division of Radiological Health Nebraska Department of Health 301 Centennial Mall South P.O. Box 95007 Lincoln, NE 68509 402-471-2168
Nevada	Stan Marshall/John Pelchat Radiological Health Section Health Division Nevada Department of Human Resources 505 East King Street, Room 202 Carson City, NV 89710 702-885-5394
New Hampshire	Belva Mohle New Hampshire Radiological Health Program Health and Welfare Building 6 Hazen Drive Concord, NH 03301-6527 603-271-4674
New Jersey	New Jersey Department of Environmental Protection 380 Scotch Road, CN-411 Trenton, NJ 08625

	800-648-0394 201-879-2062 (in state only)
New Mexico	Dave Baggett New Mexico Radiation Protection Bureau P.O. Box 968 Santa Fe, NM 87504 505-827-2938
New York	Bureau of Environmental Radiation Protection New York State Dept. of Health Empire State Plaza, Corning Tower Albany, NY 12237 518-473-3613 800-458-1158 (in state only)
North Carolina	Radiation Protection Section North Carolina Department of Human Resources 701 Barbour Drive Raleigh, NC 27603-2008 919-733-4283
North Dakota	Dale Patrick/Jeff Burgess North Dakota Dept. of Health Missouri Office Building 1200 Missouri Avenue P.O. Box 5520 Bismarck, ND 58502 701-224-2348
Ohio	Debbie Steva Radiological Health Program Ohio Department of Health 1224 Kinnear Rd., P.O. Box 118 Columbus, OH 43212 614-481-5800 800-523-4439 (in Ohio only)

Oklahoma	Coleman Smith Radiation and Special Hazards Service Oklahoma State Dept. of Health P.O. Box 53551 Oklahoma City, OK 73152 405-271-5221
Oregon	Radiation Protection Oregon Department of Human Resources 1400 SW 5th Avenue Portland, OR 97201 503-229-5797
Pennsylvania	Tim Hartman Radon Program Office PA DER Bureau of Radiation Protection 1100 Grosser Road Gilbertsville, PA 19525 800-23-RADON (in state) or 215-369-3590
Puerto Rico	David Saldana Puerto Rico Radiological Health Div. G.P.O. Call Box 70184 Rio Piedras, PR 00936 809-767-3563
Rhode Island	Roger Marinelli/Charles McMahon Division of Occupational Health and Radiation Control Rhode Island Department of Health 206 Cannon Bldg., 75 Davis Street Providence, RI 02908 401-277-2438
South Carolina	Bureau of Radiological Health South Carolina Dept. of Health and Environmental Control

	2600 Bull Street Columbia, SC 29201 803-734-4700/4631
South Dakota	Joel Smith Office of Air Quality and Solid Waste South Dakota Dept. of Water & Natural Resources Joe Foss Office Building 523 East Capital Pierre, SD 57501 605-773-3364
Tennessee	Jackie Waynick Division of Air Pollution Custom House 701 Broadway Nashville, TN 37219-5403 615-741-4634
Texas	Gary Smith Bureau of Radiation Control Texas Department of Health 1100 West 49th Street Austin, TX 78756-3189 512-835-7000
Utah	Bureau of Radiation Control Utah State Department of Health State Health Department Bldg. P.O. Box 16690 Salt Lake City, UT 84116-0690 801-538-6734
Vermont	Paul Clemens/Ray McAndliss Division of Occupational and Radiological Health Vermont Department of Health

	Administration Building 10 Baldwin Street Montpelier, VT 05602 802-828-2886
Virginia	Bureau of Radiological Health Department of Health 109 Governor Street Richmond, VA 23219 804-786-5932 800-468-0138 (in Va only)
Washington	Bruce Picket/Robert Mooney Environmental Protection Section Washington Office of Radiation Protection Thurston AirDustrial Center Building 5, LE-13 Olympia, WA 98504 206-753-5962
West Virginia	Bill Aaroe Industrial Hygiene Division West Virginia Department of Health 151 11th Avenue South Charleston, WV 25303 304-348-3526/3427
Wisconsin	State Division of Health Wisconsin Dept. of Health and Social Services 5708 Odana Road Madison, WI 53719 608-273-5180
Wyoming	Julius E. Haes Wyoming Dept. of Health and Social Services Hathway Building, 4th Floor Cheyenne, WY 82002-0710 307-777-7956

Appendix B

Nationwide Radon Detection Services

The following organizations have demonstrated their proficiency in measuring radon to the U.S. Environmental Protection Agency. All offer their services nationwide. An asterisk indicates detectors tested by CU. Prices are as quoted by suppliers for, except as noted, one detector; + indicates shipping is extra.

Activated-Charcoal Detectors

ABE Radiation Measurements Lab, RD 1, Box 214, Lenhartsville, PA 19534, 215-756-4153. $25.

*Airchek, 543 King Road. P.O. Box 2000, Arden, NC 28704. 704-684-0893. $12 +. Unit tested: *Airchek*.

Air Sciences/Overman Associates, P.O. Box 376, Bonne Terre, MO 63628. 314-358-4011. $35.

Alpha Control, 6407 Goodluck Rd., Riverdale, MD 20737. 301-474-0071. $55 a pair.

AMC Home Inspection Services, 424 Vosseller Ave., Bound Brook, NJ 08805. 201-469-6050. $75.

Amersham Corp., 2636 S. Clearbrook Drive, Arlington

Heights, IL 60005-4692. 312-593-6300. $30.

Applied Health Physics, Inc., 2986 Industrial Blvd., P.O. Box 197, Bethel Park, PA 15102. 412-563-2242. $25.

Berringer EDA Instruments, Inc., 5151 Ward Road, Wheatridge, CO 80033. 800-654-0506. $25.

CMT, Inc., 2813 Rio Vista, Emporia, KS 66801. 316-342-2760. $50.

Electro Mechanical Concepts, Inc., 130 Mountaineer Lane, W. Mifflin, PA 15122. 412-276-2272. $35.

EKS-RadTech, 1000 Herald Square, Aston, PA 19104. 215-358-9425; 800-RUALERT. $27 + a pair.

Enviroserv, P.O. Box 94, Morris Plains, NJ 07950. 201-285-1065. $40.

Geomet Technologies, Inc., 20251 Century Blvd., Germantown, MD 20874. 301-428-9898. $45 a pair.

*Health Physics Associates Ltd., 3304 Commercial Ave., Northbrook, IL 60062. 312-564-3330. $20. Unit tested: *Nodar*.

Key Technology, Inc., Box 562, Jonestown, PA 17038. 717-274-8310. $20.

Lapteff Assoc., P.O. Box 4150, Woodbridge, VA 22194. 703-491-6700. $25.

Maine State Public Health Lab, Station 12, 221 State St., Augusta, ME 04333. 207-289-2727. $18; specify "T" test.

Microbac Labs, 2411 W. 26th Street, Erie, PA 16506. 814-833-4790. $19.

Nuclear Sources & Services, Inc., 5711 Etheridge, Houston, TX 77087. 713-641-0391. $25.

O. K. Rems Corp., 174 Flock Road, Mercerville, NJ 08619. 609-588-9627. $50+.

Princeton Service Ctr., 3490 Route 1, Princeton, NJ 08540. 609-452-2037/2038. $50.

Product Analysis & Structural Test., 6800 Wales Road, Northwood, OH 43619. 419-691-8484. $17.

Radiation Protection Services, Inc., P.O. Box 2395, Darien, CT 06820. 203-324-7967. $200, cost includes "installation, inspection, etc." of three detectors.

Radiation Safety Engineering, Inc., 6105 S. Ash, Ste. 1, Tempe, AZ 85283. 602-897-9459. $20.

Radiation Service Organization, P.O. Box 1526, Laurel, MD 20707-0953. 301-953-2482. $18.

Radon Alert Detection Center, P.O. Box 323, Flourtown, PA 19031. 800-345-6348. $25.

Radon Analysts, P.O. Box 509, Livingston Manor, NY 12758. 914-292-2277, 914-439-5111. $20.

Radon Analysis, Inc., RR1, P.O. Box 561M Fox Run, Stewartsville, NJ 08886. 201-479-6086. $20.

Radon Management Company, 4332 Gingham Court, Alexandria, VA 22310. 202-547-1005. $45.

Radon Measurement and Services, 13131 W. Cedar Drive, Lakewood, CO 80228. 303-980-5086. $90 for 3; price includes installation.

*Radon Project, P.O. Box 90069, Pittsburgh, PA 15224. 412-687-3393. $12. Unit tested: *DBCA Collector*.

*Radon Testing Corp. of America, RTCA, 12 W. Main St., Elmsford, NY 10523. 800-457-2366. 914-347-5010. $30+. Unit tested: *RTCA*.

Radon Testing Service, P.O. Box 19425, Pittsburgh, PA 15213. 412-687-1533. $25.

Radon and Water Testing Service, 10806 Lafayette PC Road, Plain City, OH 43064. 614-873-8821. $13.

Radontech, Inc., 1616 Walnut Street, Suite 2200, Philadelphia, PA 19103. 215-546-7328. $175 for four; price includes installation and pickup.

Rogers & Assoc. Engineering Corp., 515 E. 4500 S., Salt Lake City, UT 84106. 801-263-1600. $18.

Ryan Nuclear Labs, P.O. Box 26687, Columbus, OH 43226-0687. 614-848-4414. $14+.

Scientific Analysis, Inc., 6012 E. Shirley Lane, Montgomery, AL 36117. 800-638-8348 in-state; 800-345-2575. $20.

Scientific Testing Associates, Ltd., Star Route, Box 124, Topping, VA 23169. 804-758-5728. $30.

Sorensen Enterprises, Inc., Radon Detection & Elimination

Service, 886 W. Hunt Road, Alcoa, TN 37701. 615-984-1376. $35.

Southern Radon Services, 1000 Johnson Ferry Road, Suite B-145, Marietta, GA 30068. 404-565-3886; 800-537-2366. $25.

TCS Industries, 4326 Crestview Road, Harrisburg, PA 17112. 717-657-7032. $19.

*Teledyne Isotopes, Inc., Environmental Analysis Dept., 50 Van Buren Ave., Westwood, NJ 07675. 201-664-7070. $50. Unit tested: *Teledyne Isotopes.*

Teledyne Isotopes Midwest Lab., 1509 Frontage Road, Northbrook, IL 60062. 312-564-0700. $25.

Alpha-Track Detectors

Radon Inspection Service, 787 E. Glen Ave., Ridgewood, NJ 07450. 201-670-8821. $35.

Radon Research Group, P.O. Box 1143, Germantown, MD 20874. 301-972-3309 or 800-544-8436; or, for out-of-state, 800-445-8436. $50.

Standor Radon Detection, 598 S. Edgewood Ave., Jacksonville, FL 32205. 904-388-0284. $25.

*Terradex Corp., 3 Science Road, Glenwood, IL 60425. 312-755-7911; or, for out-of-state, 800-528-8327. $25. Units tested: *Radtrak* and *Terradex.*

Both Detector Types

Abionics, Ltd., 12509 Kings Lake Drive, Reston, VA 22091. 703-620-3767. Charcoal or alpha-track, $65; price includes installation, consultation, pickup.

Appalachian Environmental Testing, 105 S. Union Street, Suite 326, Danville, VA 24541. 804-792-1300. Charcoal $38 a pair, alpha-track $50 a pair.

Gemini Research, Inc., P.O. Box 237, Merrifield, VA 22116. 800-272-3666, 703-941-0070. Charcoal or alpha-track, $50.

House Doctors, Inc., P.O. Box 10070, Colorado Springs, CO 80932. 303-574-6960. Charcoal $20+, alpha-track $25+.

Radon Detection Services, Inc., P.O. Box 419, Old York Road, Ringoes, NJ 08551. 201-788-3080. Charcoal or alpha-track, $50.

Radon Detection Services, Inc., P.O. Box 1195, Laurel, MD 20707. 301-725-2901. Charcoal $45, alpha-track $50.

Radon Detection Services, Inc., 1011 Brookside Road, Suite 270, P.O. Box 3309, Allentown, PA 18106. 215-481-9555. Charcoal or alpha-track, $50.

Ross Systems, Inc., Blairstown Professional Bldg., 174 Rte. 94, Blairstown, NJ 07825. 201-362-5571. Charcoal $30, alpha-track $40.

Index